COLOR
DESKTOP PRINTER TECHNOLOGY

OPTICAL SCIENCE AND ENGINEERING

Founding Editor
Brian J. Thompson
University of Rochester
Rochester, New York

COLOR
DESKTOP PRINTER
TECHNOLOGY

edited by

Noboru Ohta
Rochester Institute of Technology
Rochester, New York

Mitchell Rosen
Rochester Institute of Technology
Rochester, New York

CRC Press
Taylor & Francis Group
Boca Raton London New York

CRC Press is an imprint of the
Taylor & Francis Group, an **informa** business
A TAYLOR & FRANCIS BOOK

CRC Press
Taylor & Francis Group
6000 Broken Sound Parkway NW, Suite 300
Boca Raton, FL 33487-2742

First issued in paperback 2019

ISBN-13: 978-0-8247-5364-1 (hbk)
ISBN-13: 978-0-367-39112-6 (pbk)

Library of Congress Card Number 2005050110

Library of Congress Cataloging-in-Publication Data

Color desktop printer technology / edited by Noburu Ohta, Mitchell Rosen.
     p. cm. -- (Optical engineering ; 106)
  Includes bibliographical references and index.
  ISBN 0-8247-5364-X (alk. paper)
  1. Color computer printers. 2. Color printing. I. Ohta, Noboru. II. Rosen, Mitchell. III. Optical engineering (CRC Press) ; 106.

TK7887.7.E635 2005
681'.62--dc22   2005050110

Visit the Taylor & Francis Web site at
http://www.taylorandfrancis.com

and the CRC Press Web site at
http://www.crcpress.com

Dedication

For Alma on her early morning bike rides.

M. R.

*To Chiaki for her peerless patience
and continuing assistance.*

N. O.

Preface

Printing has been practiced since the eighth century in China, when wood blocks were used to reproduce Buddhist scriptures. As early as the 11th century, the Chinese employed movable type, although modern printing technologies trace back to Gutenberg's rediscovery of movable type in the mid-1400s. Centuries of innovation and improvement have gifted us with high-quality printing affordable to the masses. In 1984, Apple introduced the Macintosh computer with its graphical interface and paired it with the ImageWriter, a dot-matrix printer used to map screen pixels to dots on the page. The desktop printing genie was out of its bottle. The decade of the 1990s opened with the introduction of color to the desktop, first through the monitor and subsequently through printers, and it closed with full-color desktop environments ubiquitous. Today's printers are astonishingly inexpensive, robust, of incredible specifications, and with superb quality. The introduction of better, faster, cheaper models is constant. Color desktop publishing, the obvious corollary, has permeated lives and lifestyles worldwide to the point where the norms for business and personal communications surpass that which could only have been dreamed of 20 years ago.

While still an analog art prior to its partnership with computers, traditional printing included the tasks of text production, drawing figures, converting photographs into dot images, and synthesizing the pieces through printing plates. These steps were labor-intensive and time- and money-consuming, representing a considerable portion of the printing cost before inks or paper. To overcome the up-front costs, press runs needed to be high in volume. The publication of documents was, thus, expensive and required special equipment, limiting its use to an elite few. The evolution of desktop printing technology has literally eradicated these barriers. Through cost reduction, universal access, and highly intuitive desktop publishing programs, the desktop printer has truly democratized the production of high-quality paper documents.

It was foretold not long ago that computer technology would quickly yield the paperless office. To many, this seemed a fair assumption. With a preponderance of documents being generated and stored online, and with the growth of networking, e-mail, Portable Document Format (PDF), e-books, and most significantly, the World Wide Web (WWW), what could hardcopy continue to offer? It appears the answer is: quite a lot. Predicting the future is an inexact science, even when all signs point in a single direction. Alas, the opposite of the prediction has come to pass: printed page per capita has continued growth at a ferocious pace.

An all-around bird's eye view of the present art of the color desktop printer fills an important gap in the available literature. Given color printers are so thoroughly dispersed throughout the workplace and into the home, the audience for this information is enormous. Some who are interested in the subject will be considering the topic for the first time and will benefit from Part I of this book. There, Chapter 1 and Chapter 2 give a useful introduction to the basic principles of color printing and to the concepts of document and image quality. Chapter 3 goes into detail on the business and market of desktop printers, starting with a historical overview, proceeding through present day, and moving toward the future.

Chapter 4 through Chapter 7, composing Part II, are devoted to the four major color desktop printer platforms: inkjet, laser printer, thermal transfer, and film recording. The technical details will enable a deep understanding of how printers commonly work. The hope is that these chapters not only teach the current state of these technologies but also prepare the reader to place new innovations encountered in the marketplace in context. This should prolong the useful shelf-life of this book because printers, as with all computer peripherals, are faced with ever-increasing obsolescence cycle frequency.

Part III looks at contemporary and future means for digital control of color. Color management systems, particularly the International Color Consortium industry standard approach, are discussed in Chapter 8. Chapter 9 is devoted to the potential future direction of spectral printing.

A healthy dose of historical perspective permeates the chapters. This would have been appreciated by Louis Walton Sipley, author of the book *A Half Century of Color*, itself now just more than half a century old. Dr. Sipley opened his 1951 treatise with these words, as true today as they were when written more than 50 years ago:

> *In the recording of contemporary achievement in any phase of the arts and sciences there is always a background of pioneering and creative research which must be taken into consideration. If proper recognition is given to the many whose efforts have enabled the attainment of modern perfection (so-called), then those occupying today's spotlight may be expected to wear their honors with reasonable humility. This is particularly applicable to the art-science of color photography and color reproduction on the printed page which has attained its present dominant position through an evolutionary development.*

This book presents to the interested novice, as well as the imaging and printer technologist, an overview of the basics of today's color desktop printers and their history. Sufficient detail is provided while explanations of the fundamentals should ensure that all levels of reader find satisfaction.

Mitchell Rosen and Noboru Ohta
Munsell Color Science Laboratory

Center for Imaging Science

Rochester Institute of Technology

Rochester, New York

Contributors

Ross R. Allen
Hewlett Packard
Palo Alto, California, U.S.A.

Takesha Amari
Chiba University
Chiba, Japan

Jon S. Arney
Center for Imaging Science
Rochester Institute of Technology
Rochester, New York, U.S.A.

Roy S. Berns
Munsell Color Science Laboratory
Rochester Institute of Technology
Rochester, New York, U.S.A.

Yongda Chen
Munsell Color Science Laboratory
Rochester Institute of Technology
Rochester, New York, U.S.A.

Gary Dispoto
Hewlett Packard
Palo Alto, California, U.S.A.

Atsuhiro Doi
Fuji Photo Film Co., Ltd.
Kaisei-machi, Japan

Yasuji Fukase
Fuji Xerox Co., Ltd.
Sakai Nakai-Cho, Japan

Eric G. Hanson
Hewlett Packard
Palo Alto, California, U.S.A.

Akira Igarashi
Fuji Photo Film Co., Ltd.
Fujinomiya, Japan

Francisco H. Imai
Pixim Corp.
Mountain View, California, U.S.A.

Tsutomu Kimura
Fuji Photo Film Co., Ltd.
Kaisei-machi, Japan

Toshiya Kojima
Fuji Photo Film Co., Ltd.
Kaisei-machi, Japan

Kenichi Koseki
Chiba University
Chiba, Japan

Masahiro Kubo
Fuji Photo Film Co., Ltd.
Kaisei-machi, Japan

Nobuhito Matsushiro
Oki Data Corporation
Tokyo, Japan

John D. Meyer
Hewlett Packard
Palo Alto, California, U.S.A.

Nathan Moroney
Hewlett Packard
Palo Alto, California, U.S.A.

Fumio Nakaya
Document Product Company
Fuji Xerox Co., Ltd.
Sakai Nakai-cho, Japan

Frank Romano
School of Print Media
Rochester Institute of Technology
Rochester, New York, U.S.A.

Mitchell R. Rosen
Munsell Color Science Laboratory
Rochester Institute of Technology
Rochester, New York, U.S.A.

Lawrence A. Taplin
Munsell Color Science Laboratory
Rochester Institute of Technology
Rochester, New York, U.S.A.

Table of Contents

PART III   The Management of Color

Part I

Fundamentals

1 Introduction to Printing

Takesha Amari and Kenichi Koseki

CONTENTS

1.1 INTRODUCTION

Printing is defined as a permanent, graphic, visual communication medium, including all the ideas, methods, and devices used to manipulate or reproduce graphic visual messages. Generally, printing processes are accomplished by applying an inked image carrier to the substrate as it operates through a high-speed press. A letterpress plate, a lithographic plate, a gravure cylinder, and a screen used in screen printing are examples of image carriers.

For electrophotographic or inkjet printing, the data representing the images are in digital form in computer storage and the image must be created each time it is reproduced. For such digital printing technologies, image carriers are not required to transfer the printed image to the substrate. Although most people think of printing as ink on paper, printing is not limited to any particular medium. Some special printing technologies reproduce the printed image without printing ink. ISO TC130/WG1 classified the printing technology shown in Table 1.1.

This chapter explains printing processes having a concrete image carrier. These printing processes can be divided into four categories depending on the feature of the plate, namely, relief printing, planographic printing, recess printing, and through-printing, as shown in Figure 1.1.

1.2 RELIEF PRINTING

1.2.1 Historical Sketch

Relief printing is the natural and oldest printing method, using an image carrier on which the image areas are raised above the non-image areas. Letterpress printing is a kind of relief printing from movable type (Figure 1.2). Printing using movable type appeared in China and Korea in the 11th century. In 1041, a Chinese printer, Pi-Sheng, developed type characters from hardened clay. Type cast from metal in Korea was widely used in China and Japan, and by the middle 1200s, type characters were being cast in bronze. The oldest text known was printed from such type in Korea in 1397 A.D.

Half a century later in 1440, Johann Gutenberg invented a systematic letterpress technology with movable type using a wine press as a high-speed printing press. Until Gutenberg's system of separate characters for printing on a press with ink on paper, all books were laboriously handwritten by scribes. Gutenberg's most notable work was to complete a Bible with 42 lines to the page using letterpress printing (Figure 1.3). People could read the Bible without a Catholic priest. This innovation by Gutenberg was in many ways the catalyst for the Reformation by Martin Luther 50 years later. It is worth noting that this one

TABLE 1.1
ISO Printing Technology

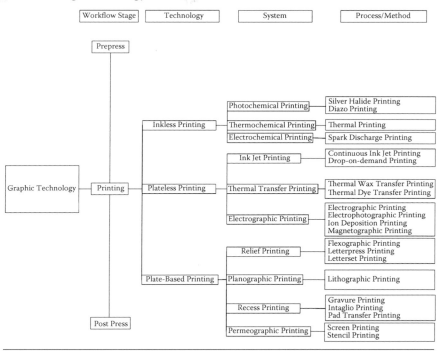

Workflow Stage	Technology	System	Process/Method
Prepress			
	Inkless Printing	Photochemical Printing	Silver Halide Printing / Diazo Printing
		Thermochemical Printing	Thermal Printing
		Electrochemical Printing	Spark Discharge Printing
	Plateless Printing	Ink Jet Printing	Continuous Ink Jet Printing / Drop-on-demand Printing
Graphic Technology → Printing		Thermal Transfer Printing	Thermal Wax Transfer Printing / Thermal Dye Transfer Printing
		Electrographic Printing	Electrographic Printing / Electrophotographic Printing / Ion Deposition Printing / Magnetographic Printing
	Plate-Based Printing	Relief Printing	Flexographic Printing / Letterpress Printing / Letterset Printing
		Planographic Printing	Lithographic Printing
		Recess Printing	Gravure Printing / Intaglio Printing / Pad Transfer Printing
Post Press		Permeographic Printing	Screen Printing / Stencil Printing

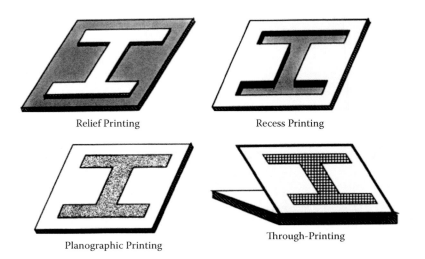

Relief Printing

Recess Printing

Planographic Printing

Through-Printing

FIGURE 1.1 Basic printing processes.

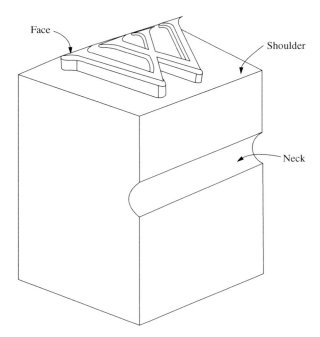

Face

Shoulder

Neck

FIGURE 1.2 Movable type.

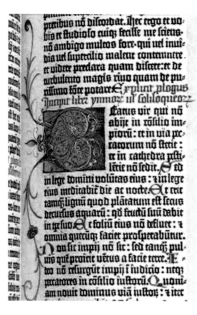

FIGURE 1.3 Gutenberg's 42-line *Bible*.

important innovation helped to terminate the Dark Age and bring about the new era of the Renaissance. High technology is not irrelevant to human life and the progress of technology does change social structure. Gutenberg's innovation and its impact on the Renaissance are good examples regarding this matter.

1.2.2 FEATURE

Any method in which the impression is taken from the raised parts of the printing surface is described as relief printing. Printing is performed by cast metal type or plates on which the image or printing areas are raised above the non-printing areas. Ink rollers touch only the top surface of the raised areas; the surrounding (non-printing) areas are lower and do not receive ink. The inked image is transferred directly to the paper.

The distinctive feature for recognizing letterpress is a heavier edge of ink around each letter (ring of ink or marginal zone), as shown in Figure 1.4. The ink tends to spread slightly from the pressure of the plate upon the printed surface. Sometimes a slight embossing (caused by denting) appears on the reverse side of the paper. The letterpress image is usually sharp and crisp.

1.2.3 PLATE MAKING

Plates used for letterpress printing can be original, duplicate, or wraparound. Original plates are usually photochemically engraved onto zinc, magnesium, or copper. A wraparound plate is a thin one-piece relief plate wrapped around the press cylinder like an offset plate. Photopolymer plates are also used as printing plates.

The oldest of the photomechanical processes, photoengraving, pertains to the production of relief printing plates for letterpress. Photoengraved plates fall into two categories: line and halftone. Generally, line and coarse screen engravings are made on zinc and magnesium, and fine screen halftone plates are made on copper.

Conventional etching is done as follows. The plate is coated with a light-sensitive coating, exposed to a negative, and then processed. The exposed coating

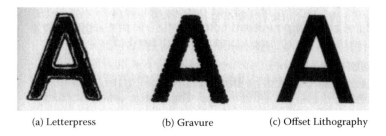

(a) Letterpress   (b) Gravure   (c) Offset Lithography

FIGURE 1.4 Characteristics of a letter printed by each printing method.

serves as a resist for protecting the image areas as the non-image areas are etched in acid baths. The main problem in conventional etching is to maintain the correct dot and line width at the proper etch depth, which is accomplished by scale compression in the negative and powderless etching.

Powderless etching can be used for zinc, magnesium, and copper plates. Zinc and magnesium use the same process. Copper uses essentially the same principles, but the chemicals and mechanism are different. The plate is prepared as in the conventional process, but a special etching machine is used. Zinc and magnesium are etched in an emulsion of dilute nitric acid, a wetting agent, and oil. During etching, the wetting agent and oil attach to the surface of the metal, forming an etch-resistant coating on the sidewalls of the etched elements, thus preventing undercutting.

In copper etching, the etchant used is ferric chloride, in which certain organic chemicals are dissolved. During etching, the additive chemicals react with the dissolved metal to form a gelatinous precipitate, which adheres to the sides of the image elements and protects them from undercutting.

Photopolymer films and plates are used in the relief printing process in which a plastic designed so it changes upon exposure to light. The photopolymer plates are used for wraparound plates and are most often used in flexography.

1.2.4 PRINTING PRESS

All printing presses have a feeding system, registration system, printing unit, and delivery system. The printing unit is the section on printing presses that furnishes the components for reproducing an inked image on the substrate under pressure.

For halftone relief printing, slight local modifications in pressure are necessary to obtain proper tone reproduction. When highlights and shadows are at the same height, the highlights exert more pressure than the shadows so that pressure must be relieved in the highlights and more pressure is added in the shadows or heavy printing areas. This procedure is known as make ready operation.

There are three types of presses: platen, flat-bed cylinder, and rotary, as shown in Figure 1.5. On platen and flat-bed cylinder presses, the type or plates are mounted on a flat surface or bed. Type and flat plates cannot be used on rotary presses, where the printing member is a cylinder, and plates must be curved.

Printing is done on sheets of paper on sheet-fed presses or on rolls of paper on web-fed presses (Figure 1.6).

A platen press carries both the paper and the type form on flat surfaces known as the platen and the bed. A flat-bed cylinder press has a moving flat bed that holds the form, and a fixed rotating impression cylinder provides the pressure. The paper, held securely to the cylinder by a set of grippers, is rolled over the form as the bed passes under the cylinder. As the bed returns to its original position, the cylinder is raised, the form is re-inked, and the printed sheet is delivered. A rotary press is the fastest and most efficient of the three types of letterpress machines and has been used mainly for long runs.

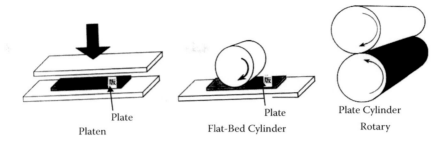

FIGURE 1.5 Printing methods used to transfer ink to the substrate in conventional printing presses.

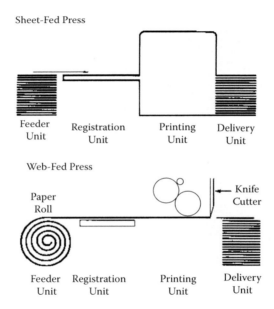

FIGURE 1.6 Schematic diagram of printing presses having four units.

1.2.5 FLEXOGRAPHY

Flexography is a high-speed web-rotary press with the relief plates made from rubber or plastic. Typical web presses for flexography are shown in Figure 1.7. The inking system consists of an ink fountain, an anilox roller system, and water- or solvent-based inks. The anilox roller is a steel or ceramic ink metering roller. Its surface is engraved with small uniform cells that carry and deposit a thin, controlled layer of ink film onto the plate (Figure 1.8). An important feature of flexographic printing is that a uniform film of ink can be printed even on rough papers because the surface of the rubber plate is sufficiently resilient so it can

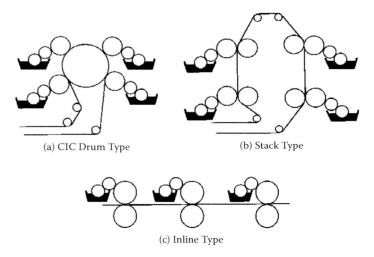

(a) CIC Drum Type   (b) Stack Type

(c) Inline Type

FIGURE 1.7 Typical web presses for flexography.

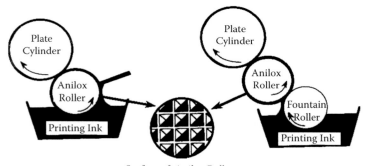

Surface of Anilox Roller

FIGURE 1.8 Ink metering system in flexo printing.

force ink into the hollows of the substrate. Using solvent type inks has made flexography possible to print on various plastic films.

1.2.6 APPLICATION

Sheet-fed letterpresses on small platen and flat-bed cylinder presses are used for short run printing, such as for letterheads, billheads, envelopes, announcements, invitations, and small advertising brochures. Larger sheet-fed letterpresses are used for general printing, such as for books, catalogs, advertising, and packaging. Web letterpresses are used for newspapers and magazines.

For nearly 400 years, Gutenberg's foundry type dominated all the printing trades. However, in the last century, especially in the last two decades, offset lithography has assumed the position of importance in the printing industry, replacing type and letterpress. Only flexography is still used in the packaging industry as relief printing.

Printing with the flexographic process includes decorated toilet tissue, bags, corrugated board, foil, hard-calendered papers, cellophane, polyethylene, and other plastic films. It is well suited for printing large areas of solid color. Inks can be overlaid to obtain high gloss and special effects. The growth of flexography parallels the expansion of the packaging industry and the development of the central impression cylinder press, new ink metering systems, and new photopolymer plates. Halftones as fine as 150 lines per inch can be printed on flexible films.

1.3 PLANOGRAPHIC PRINTING

1.3.1 HISTORICAL SKETCH

Lithography was invented by Aloys Senefelder in 1796. The printing surface was a level slab of a special form of limestone from Bavaria. The litho stone is ground flat and polished at the top surface. The image is formed by drawing or writing with a water-soluble fatty acid ink. The whole surface is treated with a weak aqueous solution of nitric acid and then etched and coated with an aqueous solution of gum arabic. After moistening the surface with water and inking with the fatty ink, the image areas accept the ink, whereas the non-image areas remain wet and reject it. Successive copies are taken by alternately dampening the stone, applying ink, and printing on paper.

1.3.2 FEATURE

Lithography is based on the principle that grease and water do not mix. On a lithographic plate the separation between the image and the non-image areas is maintained chemically because they are essentially on the same plane; the image areas must be ink receptive and refuse water, and the non-image areas must be water receptive and refuse ink (Figure 1.9). However, in reality ink and water do mix slightly. If they didn't, lithography would not be possible. If they mix too much, problems such as tinting, scumming, and bleeding occur.

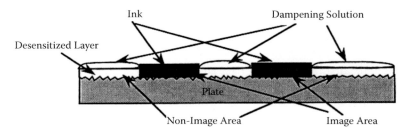

FIGURE 1.9 Schematic diagram of lithographic plate.

1.3.3 CLASSIFICATION

The planographic process is a method of printing from a flat surface. The image areas accept printing ink, and the non-image areas reject it. The best known and most widely used planographic process is lithography. Litho printing is divided into two categories: direct lithography and offset lithography. A. F. Harris invented offset lithography, observing that a sharp print was transferred to paper after accidentally first printing onto the rubber blanket of the press cyclinder. A collotype, which resembles lithography in some respects, is another form of planographic printing.

1.3.4 PLATE MAKING

On the lithographic plate, ink receptivity is achieved with inherently oleophilic (oil-loving) resins or metals such as copper or brass on the image areas. On the other hand, water receptivity of the non-image areas is usually achieved by using metals such as aluminum, chromium, or stainless steel, whose oxides are hydrophilic (water-loving). Water receptivity is maintained in plate making and storage by using natural and synthetic gums. The most widely used gum is gum arabic.

Most lithographic plates use either grained or anodized aluminum as a base. An advantage of lithographic plates, besides simplicity and low cost, is the ease of making minor corrections on the press. If corrections are extensive, however, it is more economical to make a new plate. Much of the growth in the lithographic industry in recent years may be attributed to this advantage. Automatic processors for plate making are used almost as extensively as for photography. These processors are an important factor in the use of web offset by newspapers. Some processors combine exposure with the processing and gumming, and several include coating and exposing as well. Figure 1.10 shows the schematic representation of lithographic plate making.

1.3.4.1 Presensitized Plates

Presensitized plates (PS-plates) are those in which the light-sensitive coating becomes the ink-receptive image area on the plate. Most are made from negatives. There are two types of PS-plates: additive and subtractive. On the additive plate, the ink-receptive lacquer is added to the plate during processing. On the subtractive plate, it is part of the precoating, and processing removes it from the non-printing areas. Such plates are also called surface plates. Until recently, all PS-plates were used for short or medium runs. For many years, albumin plates dominated this field, but they are now obsolete. PS-plates are currently diazo presensitized (precoated) or wipe-on (in-plant coated) for short and medium runs and prelacquered diazo presensitized and photopolymer plates for longer runs.

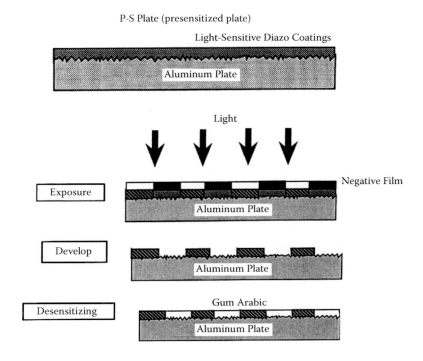

FIGURE 1.10 Planography plate-making process using P-S plate.

1.3.4.2 Waterless Plates

Waterless lithography, also called driography, is a planographic process like lithography, but it prints without dampening water. The process eliminates all of the disadvantages caused by the need for an ink–water balance in lithography but retains all the advantages of low plate costs, ease of make ready, high speed, and good print quality plus the advantages of letterpress in ease of printing and low waste. Waterless plates generally consist of ink on aluminum for the printing areas and silicone rubber for the non-image areas. Silicone rubber has a very low surface energy and, thus, resists being wet by anything, especially ink. However, under the pressure and heat of printing, ordinary litho ink has a tendency to smear over the silicone and cause scumming or toning.

1.3.4.3 Laser Plate Making

Lasers are high-energy concentrated light sources that are used for scanning and recording images at high speed. Several plate-making systems have been developed in which helium–neon (He/Ne) lasers are used to scan a pasteup, and this information is processed to expose a plate with an argon–ion laser. Because the laser can be controlled by electronic impulses, it can be operated by digital signals from a computer, thus enabling satellite and facsimile transmission of plate images to remote printing locations.

1.3.5 ELEMENTS OF OFFSET LITHOGRAPHY

Offset printing is an indirect printing method in which the inked image on a press plate is first transferred to a rubber blanket, which in turn offsets the inked impression to a press sheet. Letterpress and gravure can also be printed using the offset principle, but most lithography is printed in this way, then offset printing is a synonym of the offset lithography. Offset lithography has the following feature: the rubber printing surface conforms to irregular printing surfaces, resulting in the need for less pressure, improved print quality, and halftones of good quality on rough surfaced papers. Because the paper does not contact the lithographic plate directly, abrasive wear of the plate is reduced and the plate life increases considerably. The image on the plate is straight-reading rather than reverse-reading. Less ink is required for equal coverage, drying speeds up, and smudging and set-off are reduced.

All offset presses make one impression with each revolution of the cylinders. As shown in Figure 1.11, conventional offset presses have three printing cylinders (plate, blanket, and impression) as well as inking and dampening systems. As the plate that is clamped to the plate cylinder rotates, it comes into contact with the dampening rollers first and then with the inking rollers. The dampeners wet the plate so the non-printing area will repel ink. The inked image is then transferred to the rubber blanket, and paper is printed as it passes between the blanket and impression cylinders.

1.3.6 PRINTING PRESSES

There are two types of offset lithography presses: the sheet-fed press and the web press (Figure 1.12). A sheet-fed press feeds and prints on individual sheets of paper. A printing press that prints on both sides of the paper using the blanket-to-blanket principle in one pass is called a perfecting press. Some sheet-fed presses are designed as perfecting presses. A web press is a rotary press that prints on a continuous web of paper fed from a roll and threaded through the press. Web presses are gradually replacing the sheet-fed press.

Sheet-fed printing has the following advantages: a large number of sheet or format sizes can be printed on the same press and waste sheets can be used during make ready, so good paper is not spoiled while getting position or color up for running. Sheet-fed lithography is used for printing advertising, books, catalogs, greeting cards, posters, packaging, decalcomanias, and art reproduction.

Much of the growth in the lithographic industry in recent years can be attributed to web offset, which is used to produce newspapers, magazines, business forms, computer letters, mail order catalogs, gift wrappings, books, and a variety of commercial printing. The latest innovations in web offset are in

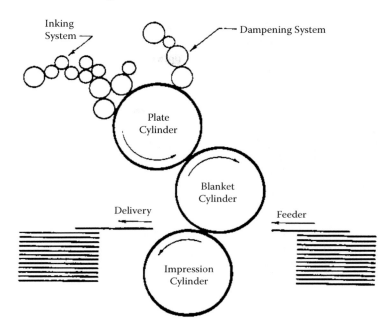

FIGURE 1.11 Elements of offset lithography press.

common impression, keyless tower, and in-line design, which make web offset competitive with gravure in long run printing. Speed is the main advantage of web offset. Speeds of 300 meters per minute are common, and new presses are designed for speeds of 500 meters per minute and faster. Most web offset presses have in-line folders where various combinations of folds convert the web into folded signatures. Other in-line operations that can be performed on-press include paste binding, perforating, numbering, rotary sheeting, and slitting. All of these make web offset very flexible, and all are done while the presses are running at high speeds, up to four times faster than sheet-fed presses.

The main disadvantage of web offset and web letterpress is that they have a fixed cut-off (i.e., all sheets cut off at the same length).

There are four types of web offset presses:

1. The blanket-to-blanket press has no impression cylinders (Figure 1.13). The blanket cylinder of one unit acts as the impression cylinder for the other and vice versa. Each printing unit has two plates and two blanket cylinders. The paper is printed on both sides at the same time as it passes between the two blanket cylinders.
2. The in-line open press is similar to a sheet-fed offset press, except the cylinder gap is very narrow. Grippers and transfer cylinders are eliminated. Each unit prints one color on one side; additional units are required for additional colors. To print the reverse side, the web is turned over between the printing units by means of turning bars, which

Sheet-Fed Press

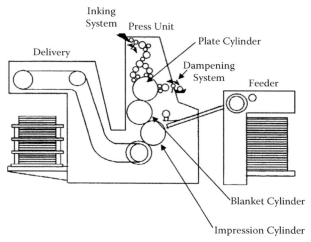

Web Press

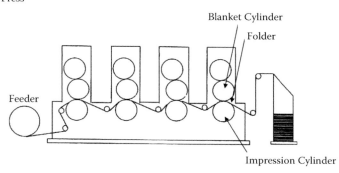

FIGURE 1.12 Sheet-fed and web-fed offset printing presses.

expose the unprinted side of the web to the remaining printing units. This type of press is used extensively for printing business forms.

3. The drum or common impression cylinder (CIC) press has all the blanket cylinders grouped around a large common impression cylinder (Figure 1.14). Some multi-color web presses designed as two to five colors are printed in rapid succession on one side, after which the web is dried and turned, and printed on the reverse side.

4. Keyless tower printing system: In ink supplying systems, anilox rollers are used to meter the ink quantity instead of the ink adjusting key and the printing ink emulsifies the dampening solution. Each printing unit is designed as perfecting and stacks vertically. Usually, this printing system is used in multicolor news printing (Figure 1.15).

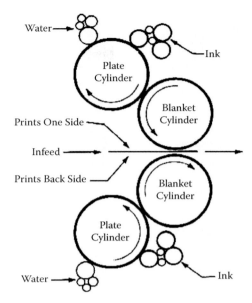

FIGURE 1.13 Blanket-to-blanket–type perfecting press.

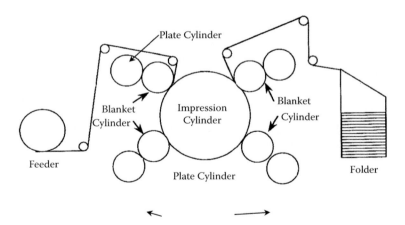

FIGURE 1.14 Common impression cylinder web press.

1.3.7 DAMPENING SYSTEMS

The conventional dampening system on offset presses transfers the dampening solution directly to the plate. In the Dahlgren type of direct-feed dampening system, the fountain solution, containing up to 25% alcohol, is metered to the plate through the inking system, or it can be applied directly to the plate, as in other systems. In general, this type of dampening system uses less water and

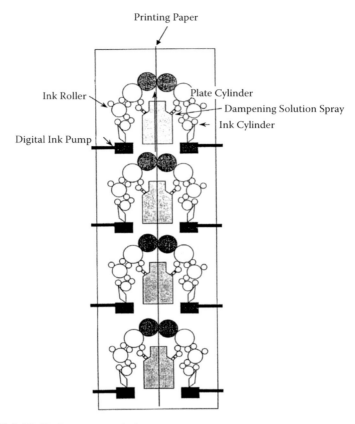

FIGURE 1.15 Keyless tower printing system.

reduces paper waste at the start-up of the press. Because of the cost of isopropyl alcohol and regulation of volatile organic compounds (VOCs), a number of new fountain solutions have been developed to reduce or replace the alcohol in this type of dampening system.

1.3.8 INKING SYSTEMS

Inking systems are designed to transport ink from the ink fountain to the printing plate. All systems use composition rollers. Some have plastic coated rollers and others have copper-plated steel rollers to prevent stripping of the ink on the distributors. Some inking systems, especially on web presses, are water cooled. The large number of rollers is needed in inking systems because litho inks are thixotropic materials, and sufficient kneading by the rollers is necessary for good fluidity. With old albumin and zinc plates, which require much water for damp-ening, a large roller surface area is needed to evaporate the water from the ink to keep it from waterlogging and emulsifying.

1.3.9 Troubles in Lithography

Damages to the planographic plate and the lack of water–ink balance are characteristic of the troubles in lithography. Scumming and greasing damage the planographic plate and tinting, emulsifying, bleeding, washing, and spreading are due to the lack of water–ink balance. The non-imaging area on the litho plate is protected by a desensitizing layer to refuse the ink; however, once the layer is damaged by foreign material or grease, these areas begin to take ink. This phenomenon is called scumming. Tinting occurs when a lithographic ink picks up too much dampening solution and prints a weak snowflake pattern. In extreme cases, the ink actually emulsifies in the dampening solution, causing an overall tint to quickly appear on the unprinted areas of the sheet.

1.4 RECESS PRINTING

1.4.1 Classification

Recess printing is a method of taking impressions from recesses engraved or etched in the printing plate or cylinder. The most important application is rotary photogravure printing, but many other methods in intaglio are still used in important fields such as bank notes and stock certificates. Schematic classification of recess printing is shown in Figure 1.16.

1.4.2 Gravure

In rotary photogravure printing, image areas consist of cells or wells etched into a copper cylinder or wraparound plate, and the cylinder or plate surface represents

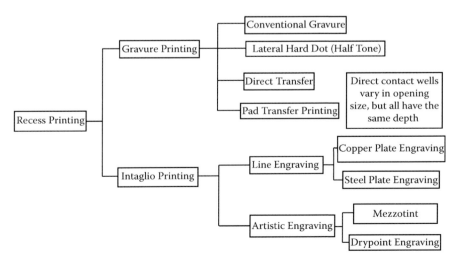

FIGURE 1.16 Schematic classification of recess printing.

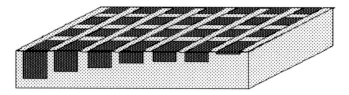

FIGURE 1.17 Conventional gravure plate.

the non-printing areas, as shown in Figure 1.17. The basic elements of gravure printing are shown in Figure 1.18. The plate cylinder rotates in a bath of ink. The excess is wiped off the surface by a flexible steel doctor blade. The ink remaining in the thousands of recessed cells forms the image by direct transfer to the paper as it passes between the plate cylinder and the impression cylinder. Gravure printing is considered to be excellent for reproducing pictures, but high plate-making expense usually limits its use to long runs. A distinctive feature for recognizing gravure is that the entire image is screened — not only halftone images but also type and line drawings.

1.4.2.1 Historical Sketch

The history of gravure printing began with the work of creative artists during the Italian Renaissance in the 1300s. The recognized inventor of modern gravure

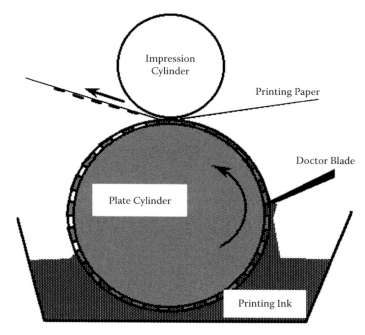

FIGURE 1.18 Basic elements of gravure printing.

printing is Karl Kliche. He began experimenting with photographic copper etching in 1875. He made the revolutionary move from flat printing plates to cylinder forms. He developed the first doctor blade and even designed a method of printing color on a web press. He originated the term *rotogravure* for printing from a cylinder.

1.4.2.2 Plate Making

1.4.2.2.1 *Chemical Method*

In conventional gravure, the image is transferred to the copper cylinder by the use of a sensitized gelatin transfer medium known as carbon tissue. The process of plate making of gravure is shown in Figure 1.19. The carbon tissue is first exposed in contact with an overall gravure screen, which is usually 150 lines per inch, virtually invisible to the naked eye (Figure 1.20). The gravure screen serves a purely mechanical purpose and, unlike other processes, has nothing to do with producing the tones of the picture. It merely provides the partitions or walls of the cells etched into the cylinder to form a surface of uniform height for the doctor blade to ride on. Then the continuous-tone positives are exposed in contact with the carbon tissue. In the highlight tones in the positives, where light passes through freely, the gelatin on the carbon tissue becomes proportionately harder with light intensity. The exposed carbon tissue is positioned on the copper plate or cylinder with precision machines. After removal of paper backing, the tissue is developed by hot water, leaving gelatin of varying thickness in the square dot areas between the hardened screen lines. The etching is done in stages using solutions of ferric chloride at varying concentration levels. Photographic resists are being developed to replace the carbon tissue. They are more stable, easier to use, and can be stored for a longer period. Conventional gravure is used for high-quality illustrations but mainly for short runs because of doctor blade wear of the shallow highlight dots.

Gravure cylinders are chromium plated for long runs. On very long runs, the chromium is worn off by friction of the doctor blade over the cylinder. In such cases the chromium-plated layer on the cylinder is removed, rechromed, and replaced in the press for continuing the run.

1.4.2.2.2 *Electromechanical Engraving*

Another method of electromechanical engraving is the Helioklischograph. In this equipment, positives or negatives of the copy made on a special opaque white plastic are scanned as the cylinder is being engraved electromechanically by special diamond styli, as shown in Figure 1.21. Research is being done on the use of electron-beam and laser etching of the copper cylinder, which are considerably faster than electromechanical engraving. The Lasergravure process by Crosfield uses a plastic-coated, copper-plated cylinder in which helical grooves with variable depths and cross bridges are etched with a 100-W CO_2 laser driven by a scanner or other front end system.

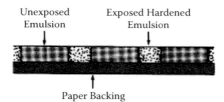

(a) Side view of carbon tissue after exposure to a cross-line screen

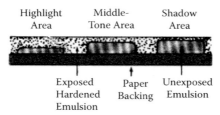

(b) Side view of carbon tissue after exposure to a continuous-tone film positive

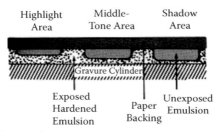

(c) Side view of carbon tissue after transfer on the gravure cylinder

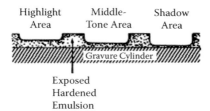

(d) Side view of carbon tissue after development by hot water

FIGURE 1.19 Conventional gravure plate-making process.

1.4.2.2.3 The Variable Area–Variable Depth Plate (Figure 1.22)

The process for long run periodical printing differs from the conventional gravure process just described in that the size of the cells as well as the depth varies to produce more durable tones in publication printing. The cells of lighter areas are smaller but deeper. In this process, continuous-tone positives and resists are used

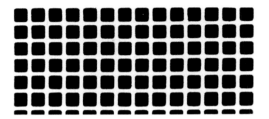

FIGURE 1.20 Cross-line gravure screen.

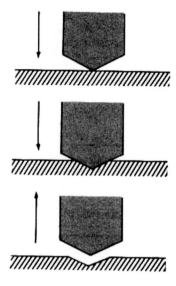

FIGURE 1.21 Electromechanical engraving with special diamond styli.

as in conventional gravure but, instead of the overall screen, a special halftone positive is used. In the direct transfer or variable area method, an acid-resistant, light-sensitive coating is first applied to the copper. The screened positive is wrapped around the cylinder and exposed directly to it by a strong light source through a narrow slit as the cylinder turns. The cylinder is then developed, and the coating, which has not been exposed by light, is removed. The cylinder is etched, producing image elements that vary in area but not in depth, so the number of tones is limited. This method is used widely in packaging and textile printing.

1.4.2.2.4 Pad Transfer Printing

This process is classified to the indirect recess printing process whose inked image areas are transferred from engraved steel or plastic forms onto substrates by means of a flexible silicon pad that adapts itself to the surface of irregularly shaped objects.

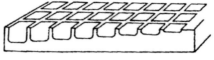

Conventional Gravure Plate
Variable Depth, Constant Area

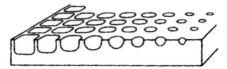

Halftone Gravure Plate (Dalgian Method)
Variable Depth, Variable Area

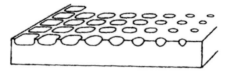

Halftone Gravure Plate (Direct Method)
Constant Depth, Variable Area

FIGURE 1.22 Structure of gravure plate.

1.4.2.3 Printing Process

Almost all rotogravure presses are designed for multicolor and web-fed gravure
presses (Figure 1.23). Rotogravure printing units consist of a printing cylinder,
an impression cylinder, and an inking system. Ink is applied to the printing
cylinder by an ink roll or spray, and the excess is removed by a doctor blade and
returned to the ink fountain. The impression cylinder is covered with a rubber
composition that presses the paper into contact with the ink in the tiny cells of

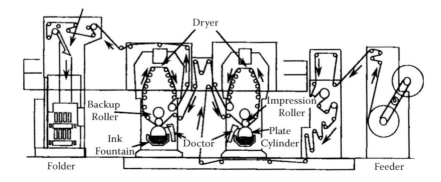

FIGURE 1.23 Multicolor web-fed gravure presses.

the printing surface. Gravure inks are volatile and dry almost instantly. Hot air dryers are used between printing units to speed up drying. Therefore, in color printing each succeeding color is printed on a dry color, rather than on one that is still wet as in letterpress and offset. For color printing, presses use photoelectric cells for automatic register control. Cylinders are chromium plated for press runs of a million or more. When the chromium starts to wear, it is stripped off and the cylinder is rechromed. One disadvantage of gravure for publication printing has been the inability to change pages on the cylinder. Wraparound printing cylinder segments have been introduced that give gravure this added capability.

Sheet-fed gravure presses operate on the same rotary principle as rotogravure. The preparatory work is identical. The image is etched flat on a flexible sheet of copper, which is then clamped around the plate cylinder of the press. Sheet-fed gravure is primarily used for short runs and press proofing. Because of the high quality and plate-making expense, it is used for art, photographic reproductions, and prestige printing such as annual reports. In packaging, sheet-fed gravure presses are used for printing new packages for market testing.

Offset gravure has been used for printing wood grains and in packaging. In this application, a converted flexographic press is used. The anilox roller is replaced by a gravure cylinder and doctor blade for printing the image and the plate cylinder of the flexographic press is covered with a solid rubber plate.

1.4.2.4 Application

Gravure printing is used in packaging for quality color printing on transparent and flexible films in which any cut-off length is possible by changing the size of the printing cylinder. Gravure printing is also used for printing cartons, including die-cutting and embossing, which can be done in-line on the press. Most long run magazines and mail order catalogs are printed by gravure. Among the specialties printed by gravure are vinyl floor coverings, upholstery and other textile materials, pressure-sensitive wall coverings, plastic laminates, imitation wood grains, tax and postage stamps, and long run heat transfer patterns.

1.4.3 INTAGLIO

As shown in Figure 1.16, intaglio printing is divided into two categories: line engraving and artistic engraving. Copperplate engraving and steel engraving belong to line engraving, and drypoint engraving and mezzotint engraving belong to artistic engraving. Drypoint engraving is produced by direct drawing using an engraving needle without etching solution. Mezzotint is engraving produced on a roughened copper plate. The highlight and midtone areas are hand-scraped to reduce ink retention, and shadow areas are burnished to strengthen ink retention. The printing ink used in intaglio printing is solidlike, whereas in gravure printing, solvent-type ink is used. This is the distinct point in the difference between gravure printing and intaglio printing.

1.4.3.1 Line Engraving

1.4.3.1.1 Copperplate Engraving

Engraving is a highly skilled art in which lines of varying depth and width are cut into metal plate with engraving tools. The plates are printed on a copper plate press, which has a flat iron bed between two rollers. The plate, after warming to soften the ink, is inked with stiff copper plate ink, and the excess is wiped off the surface with wiping canvas and finally with the palm of the hand. The inked plate is laid on the iron bed of the press and the paper (previously damped to soften the fibers) is placed in position over it and backed with layers of printer's blanket. When rolling pressure is applied, the paper is forced into intimate contact with the plate to receive the ink.

1.4.3.1.2 Steel-Die Engraving

Steel-die engraving is line engraving in which the die is hand or machine cut, or chemically etched to hold ink. The plate is inked so that all sub-surfaces are filled with ink. Then the surface is wiped clean, leaving ink only in the depressed (or sunken) areas of the plate. The paper is slightly moistened and forced against the plate with tremendous pressure, drawing the ink from the depressed areas. This produces the characteristic embossed surface, with a slightly indented impression on the back of the paper.

1.4.3.1.3 Mezzotint

In mezzotint engraving, small recesses are formed over the whole plate surface with a serrated-edge rocking tool that throws up a burr of displaced metal. Gradation of tone is obtained by using scrapers and burnishers to remove the burr and reduce the depth of the recesses. This art process is sometimes used for color printing, the different colored inks being applied locally to the plate for printing at one impression.

1.4.3.2 Application

Copper plates are used for short runs of one-time use (invitations and announcements). For longer or repeat runs such as letterheads, envelopes, greeting cards, stamps, bank notes, and stock certificates, chromium-plated copper or steel plates are used in a die-stamping press. Line engraving is used to produce plates for some forms of maps and charts, for producing litho transfers, and for printing invitation cards, visiting cards, and similar work.

1.5 THROUGH-PRINTING

Compared to other printing technology, through-printing has a characteristic of an industrial art object. Practically, screen printing has been used for art prints, posters, decalcomania transfers, greeting cards, menus, program covers, and wallpaper. Screen printing is important in the printing of textiles such as tablecloths, shower curtains, and draperies. On the other hand, the remarkable development of the electronic industry in the 1970s depended on screen printing technology. Screen-stencil technology is indispensable to large-scale integration (LSI) manufacture. Now, screen printing is a basic technology for integrated circuit (IC) industries.

1.5.1 FEATURE

In 1907, Samual Simon of Manchester, England, got the idea for screen printing from the traditional dying process of Yuzen-zome in Kyoto, Japan. The screen printing process originated as a method of printing from stencils supported on fabric stretched on a frame. Screens made with perforated metal and other materials are sometimes used.

Formerly known as silk screen, this method employs a porous screen of fine silk, nylon, dacron, or stainless steel mounted on a frame. Figure 1.24 shows microphotography of the screen fabric. A stencil is produced on the screen, either manually or photomechanically, and the non-printing areas are protected by the stencil. Printing is done on paper or other substrate under the screen by applying ink with a paintlike consistency to the screen, spreading and forcing it through the fine mesh openings with a rubber squeegee. At the same time, the squeegee scratches out the excess ink from the surface of screen fabric. The ink pressed through the open image areas of the screen form the inked image on the substrate. This printing process is shown in Figure 1.25.

1.5.2 CLASSIFICATION

As the name suggests, through-printing is the method of obtaining an inked image through the openings in a screen fabric. The basic concept of through printing is simple and is based on the idea of a stencil. By taking a piece of paper, drawing some outline or sketch of an object on it, and then cutting out the sketch, we can get a stencil. Two important printing methods are stencil duplicating and screen printing.

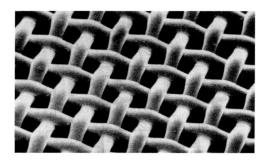

FIGURE 1.24 Micrograph of screen fabric.

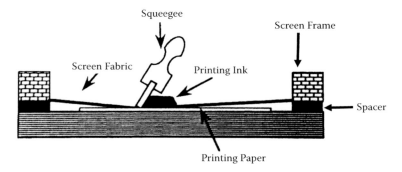

FIGURE 1.25 Screen printing by hand.

Generally, stencil duplicating is performed by off-contact printing, and screen printing is done by on-contact printing. In stencil duplicating, after painting, spraying, or depositing ink on the stencil, ink passes through perforations in a stencil master to form a printed image on a substrate. After the image is transferred, the frame must be hinged up and the stock carefully peeled away.

As shown in Figure 1.25, in screen printing, the screen is slightly raised away from the printing material by small shims under the hinge and frame. With this technique, the screen touches the paper stock only while the squeegee passes over the screen. Once the image is transferred across the squeegee line, the screen snaps back away from the substrate. Off-contact printing helps keep the press sheet from sticking to the screen and usually prevents image smearing. Figure 1.26 shows the schematic diagram of ink transfer in the screen printing process.

1.5.3 PLATE

There are many methods of making plates for screen printing. As previously mentioned, the screen consists of a porous material, and the printed image is produced by blocking unwanted holes or pores of the screen. Early screen plates were made manually by painting the image on silk fabric mounted on a wooden frame. Masking materials were used to block out unwanted areas. Today, both hand-cut stencils and photomechanical means are used. In the photomechanical method the screen is coated with a light-sensitive emulsion; exposure is made through a screened film.

1.5.4 PRINTING PROCESS

Some screen printing is done by hand with very simple equipment consisting of a table, screen frame, and squeegee, as shown in Figure 1.25. Most commercial screen printing, however, is done on power-operated presses. There are both roll-fed and sheet-fed presses, with hot air dryers, which run at speeds up to 400 feet per minute or more than 5000 impressions per hour.

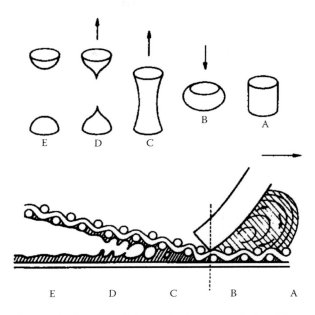

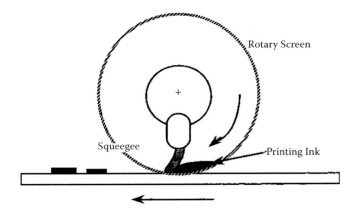

FIGURE 1.26 Schematic diagram of ink transfer in screen printing. Ink profile in screen opening at corresponding position as indicated.

There are two types of power-operated presses. One type uses flat screens, which require an intermittent motion as each screen is printed. Butts and overlaps require close register, which limits running speed. The latest type uses rotary screens with the squeegee mounted inside the cylinder and the ink pumped in automatically (Figure 1.27). These presses are continuous running, are fast, and print continuous patterns with little difficulty. The amount of ink applied by screen printing is far greater than in letterpress, lithography, or gravure, which accounts for some of the

FIGURE 1.27 Diagram of a rotary screen printing unit.

unusual effects in screen printing. Because of the heavy ink film, the sheets must be racked separately until dry or passed through a heated tunnel or drier before they can be stacked safely without smudging or set-off. Ultraviolet curing ink has effective drying and is helping to promote greater use of screen printing.

1.5.5 APPLICATION

Versatility is the principal advantage of screen printing. Screen printing prints on almost any surface, and both line and halftone work can be printed. Any surface can be printed such as wood, glass, metal, plastic, fabric, and cork, in any shape or design, any thickness, and any size. In advertising, screen printing is used for banners, decals, posters, 24-sheet billboards, car cards, counter displays, menu covers, and other items. Heavy paperboards can be printed, eliminating costly mounting. Wallpapers and draperies are printed because of the depth of colors afforded, especially in the short run custom designs of interior decorators. There are many other specialty uses for screen printing, such as decorating melamine plastic sheets before lamination and the printing of electronic circuit boards.

2 Image Quality of Printed Text and Images

Jon S. Arney

CONTENTS

2.1 THE MEANINGS OF IMAGE QUALITY

This chapter focuses primarily on the characteristics of printer systems that have a significant impact on the visual quality of printed images. Characteristics such as the absorption spectrum of printing inks, the gamma of photoconductors, and satellite drops in inkjet are examples of system characteristics that impact image quality. The relationship between system characteristics of this kind and the ultimate quality of a printed image is of considerable practical importance, but it is not always easy to quantify the connection. The term *image quality* is used in many different ways, so before considering printers and printer characteristics, it is helpful first to consider the various ways the term image quality is used.

2.1.1 THE IMAGE QUALITY CIRCLE

Many different technologies have been developed into commercially successful printing systems. The traditional printing technologies described in Chapter 1 and the newer digital techniques described in Part II of this text differ significantly in the way in which colorant is applied to paper. However, all these different technologies have two attributes in common. They all produce printed images of high quality, and they do so at a low price. Many other technologies for putting colorant on paper have been developed and described in the technical literature, but only those capable of quality and economy are commercially successful. Thus, both quality and economy are topics essential for anyone in the printing industry to understand.

    The meaning of the term *economy* is easy to understand quantitatively, but image quality is a more difficult term to express in a quantitative way. The difficulty is the subjective way in which customers use and describe the quality of printed images. Nevertheless, if a printing process does not provide the customer with enough quality, regardless of the ability to understand and quantify it, the printing process will not survive commercially. The overall quality of an image is therefore measured indirectly by the commercial success of the printing process.

    To predict image quality, and thus commercial success, other parameters are measured that are believed to serve as useful quality metrics. The relationship between overall image quality and these other kinds of metrics can be represented as shown in the Engeldrum Image Quality Circle shown in Figure 2.1. The different types of image quality metrics are represented by the boxes, and the relationships among the types of metrics are represented by the ovals.

2.1.2 IMAGE QUALITY METRICS

The overall quality of an image, represented by box (A) at the top of Figure 2.1, is generally believed to be made up of a group of perceptual attributes represented

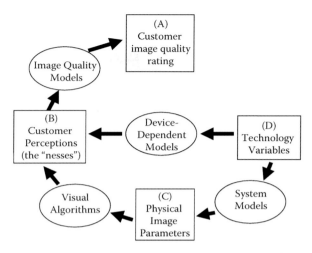

FIGURE 2.1 The image quality circle. (Modified from P.G. Engeldrum, *Psychometric Scaling: A Toolkit for Imaging Systems Development*, Imcotek Press, Winchester, MA, 2000.)

by box (B) on the left side. Examples of these component attributes are colorfulness, sharpness, graininess, saturation, and crispness. These individual attributes, sometimes called the "nesses," can be quantified experimentally with well-known techniques of psychophysical testing.[1] Thus, these attributes can be expressed as unambiguous, quantitatively measurable metrics. For example, a group of observers might be asked to judge the relative degree of graininess, colorfulness, or sharpness of a set of test images. Appropriately designed testing can lead to numerical scales of graininess, colorfulness, or any of the other "nesses."[1]

The set of test images can also be characterized with a variety of instrumental measurements. These measurements provide the physical image parameters shown in box (C) at the bottom of Figure 2.1. Examples might include the reflectance spectrum, the noise power spectrum, and the image histogram. From these functions, individual metrics such as L*, a*, contrast, resolution, and granularity can be extracted. These physical image parameters are often used as predictors of the "nesses" in Figure 2.1.

Box (D) on the right side of Figure 2.1 represents the technology variables, also called the printer device parameters. These are the parameters that describe the operation of the printing device, and they ultimately control the characteristics of the image in the box (C) at the bottom of Figure 2.1. Examples of device variables and functions containing device variables include the device modulation transfer function (MTF), the paper roughness, the colorant concentration, and the device gamma. These are variables over which the printer manufacturer and the printing service company have some control, and it is common to search for the optimum values of the variables that will ensure the highest quality of printed image. The relationship between the variables of the printing device (box (D) of Figure 2.1) and the overall quality of the image (box (D) of Figure 2.1) and the relationship between any two boxes in Figure 2.1 are referred to as image quality models. The different types of image quality models are represented as ovals in Figure 2.1.

2.1.3 IMAGE QUALITY MODELING

Each oval in Figure 2.1 represents a different type of image quality model for estimating parameters in one box based on the parameters of another box. Each type of model in Figure 2.1 has its usefulness.

Device-Dependent Models (D→B): Modeling the relationship between the nesses and the technology variables has been called device-independent modeling.[2,3] This type of modeling is particularly important when a printer must be adjusted to provide the highest print quality in the shortest possible time. A powerful technique for constructing such a model is to apply an empirical technique called statistical experimental design[4] in which all of the relevant variables of the printing system (ink density, photoconductor gamma, paper type, print speed, toner fusing temperature, etc.) are varied in all relevant combinations. For each combination of variables, a print is generated and evaluated by observers in a psychophysical experiment to determine a set of perceptual metrics (sharpness, colorfulness, graininess, etc.). Then a statistical optimization is carried out to determine the values of the printer variables that produce the best set of perceptual values, or nesses. This sort of device-dependent modeling is a powerful way to calibrate a printer for the best quality in the shortest amount of time. However, it does not provide much guidance in understanding the underlying nature of image quality or how it relates mechanistically to the printing process. Thus, other kinds of image quality modeling are also important.

System Models (D→C): The properties of the printing system directly control the properties of the image it produces. Thus, for example, a change in the gamma of a photoconductor in a laser printer can change the resolution of an image printed with the printer. It is important for product development specialists to understand the relationship between the technology variables and the properties of the printed images. Models relating these metrics are also useful for trouble-shooting difficulties with a printing system.

Visual Models (C→B): Images can be characterized by a variety of instrumental techniques. Common examples are reflectance spectroscopy, colorimetry, histogram analysis, microdensitometry, gloss, and goniophotometry. Experimental metrics derived from such measurements are often called objective quality metrics and are often correlated with the results of psychophysical analysis. This kind of modeling does not depend on the properties of an imaging device and involves only the measurable properties of the image. For this reason, models of this kind are sometimes called "device-independent models."[4]

Combining a visual model with a system model provides a quantitative, mechanistic link between the perceptual nesses of an image and the underlying technology variables that control the image-producing device. The advantages of such a two-step model are clear, but achieving good, predictive models of this sort is challenging and time consuming.

Image Quality Models (B→A): Modeling the overall quality of an image based on the collection of perceptual nesses in box (B) of Figure 2.1 is a challenge that is still a major topic of research among image scientists, experimental psychologists, and market analysts. As yet, no generally satisfactory solution has been developed, and empirical trial-and-error is still the principal technique employed.

The remainder of this chapter focuses on the technology variables (Figure 2.1, box (D)) of printing systems and their impact on the so-called objective image quality metrics of printed images (Figure 2.1, box (C)).

2.2 THE PRINTING OF TEXT

Image quality requirements for document images containing only alphanumeric characters are fewer than the requirements for pictorial image printing. However, achieving a high level of text quality is not a trivial matter, and printer characteristics that influence text quality also play key roles in pictorial image quality. Moreover, changes in the printing system that favor text quality often disfavor pictorial quality, leading to a need for text/pictorial tradeoff. Thus, before exploring color image quality, it is essential first to examine text image quality.

In general, text characters must have high density relative to the paper and must have sharp edges, serifs and other fine details must be reproduced repeatably, gloss differences between printed and not-printed regions should be minimized, and there should be no stray colorant between the characters. These requirements seem self-evident, but the achievement of these requirements is not a trivial matter. Moreover, the competitive drive toward higher image quality has led to an overall increase in consumer expectations for text image quality. During the 1970s, the benchmark for office text quality was the impact typewriter. Improvements in ink ribbons and the introduction of the electric typewriter raised the bar for text image quality, and the best quality achievable with an electric typewriter during the 1970s became known as a letter-quality text image. The development of desktop computing and printing during the 1980s raised the bar even higher. The early printers were dot-matrix, impact printers capable of letter-quality printing, but the introduction of inkjet and laser printers quickly demonstrated the ability to produce letter-quality prints. By the end of the 1980s, desktop printers were available for routine office use and produced prints that were nearly indistinguishable from many offset lithographic presses. The term *letter-quality* is no longer used, and offset-litho quality is now the benchmark for desktop document printing.

2.2.1 PRINTER RESOLVING POWER AND ADDRESSABILITY

Marketing literature for printers and printing systems often describes resolution in terms of dots per inch (dpi) or dots per millimeter (dpm). However, this is often misleading. The dpi and dpm metrics are really the addressability of the

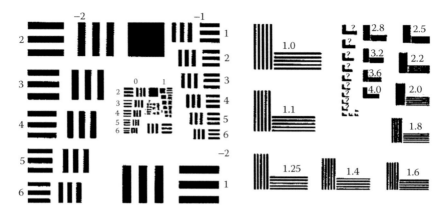

FIGURE 2.2 USAF (left) and NBS (right) test charts.

system. This is the maximum number of locations per unit length on the substrate paper where the printer can deliver a unit of colorant (dot). Although addressability is an important metric for a printing system, it is not equivalent to resolution.

Resolution is generally measured quantitatively by an index of resolving power. This is done by printing test characters or bar patterns of smaller and smaller sizes. The smallest pattern size with bars that can be unambiguously distinguished by someone using an appropriate optical magnification device is used as the index of resolving power.[5] Numerous test patterns and test procedures have been published for quantifying resolution, and Figure 2.2 illustrates two very common ones. The United States Air Force (USAF) resolution test chart, shown on the left, was originally developed during the 1940s to characterize aerial reconnaissance film,[6] and the National Bureau of Standards (NBS) Resolution Test Chart was developed to measure the resolution of optical microscopes.[7] These test charts and others have been adapted to the characterization of printing systems, and a survey of current technical literature is recommended for anyone involved in this kind of testing.[1,5,8] Test charts such as these are available commercially, both for image capture (cameras, microscopes, scanners, etc.) and for image output. Digital files with resolution charts, for example, are useful for characterizing the resolution of digital printers.

Printer addressability certainly has an impact on printer resolution. However, as illustrated in Figure 2.3, addressability is not the only factor controlling resolution. The grid of addressable points in Figure 2.3 is illustrated with simulated ink dots of one half, one, two, and four times the distance between addressable locations, and it is clear that large dots, or dots that spread significantly on the printed paper, will not produce images with the same resolution as can be achieved with small dots. Thus, dot quality is as important to printed image resolution as is the dpi addressability of a printing system.

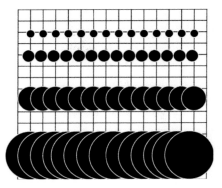

FIGURE 2.3 Grid of addressable points with dots at consecutive locations. Dots are illustrated with diameters that are half, one, two, and four times the distance between addressable points.

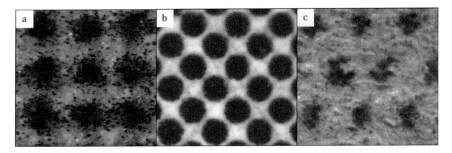

FIGURE 2.4 Examples of different dot shapes produced by different inks, papers, and printing technologies. (a) 300-dpi electrophotographic laser printer on plain paper. (b) 150-dpi offset lithography on magazine quality paper. (c) 300-dpi inkjet drops forming 4 × 4 dot clusters at 37.5 lines per inch on plain paper.

Printing devices also can have dpi addressability that is different in the horizontal and vertical directions, and ink dots can be of many shapes other than round. Moreover, the edges of dots are not always clearly defined, as illustrated in Figure 2.4. These kinds of effects are quite important both to image resolution and to pictorial color reproduction. This is the reason for both horizontal and vertical line patterns in the test charts illustrated in Figure 2.2.

In some technologies, the resolution of the image can be higher than the dots per inch of the printing system. This is true for the traditional photographic process for making halftone printing plates for offset lithographic printing, as illustrated in Figure 2.5 (see also Section 2.4.1). The photographic process controls not only the sizes of halftone dots but also the shapes. This results in partial dots that align along sharp lines, thus producing images with a higher resolution than the nominal dpi dot spacing would suggest.[9]

FIGURE 2.5 Photographic process halftoning produces printing dots with higher resolution than the dot pitch. (Courtesy of Frank Cost.)

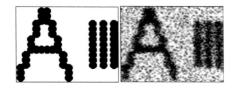

FIGURE 2.6 Illustration of line jags produced by discrete dots on a printing device. The right-hand image is identical but with significant noise and blur added in a simulated printing process.

As illustrated in Figure 2.4, there may be a variety of technological factors that spread ink dots and lead to microscopic blurring of the image structure. Although this can only decrease resolution of an image, it is not always true that it decreases image quality. There are cases, as will be described subsequently, in which micro-blurring can lead to better color reproduction. As shown in Figure 2.6, some amount of blurring can reduce the visual impact of a digital printing artifact, sometimes called line jags.

2.2.2 INK DENSITY

For a printing process to produce high-quality text, the text characters must be very dark. Mechanical typewriters of the mid 20th century print text that is much lighter than the text produced by an offset web press. The distinction between letter-quality text and offset-quality text is primarily the darkness of the printed characters. The lightness and darkness of a printed image can be measured accurately by comparing the printed image to a photographic step tablet, as illustrated in Figure 2.7. A photographic step tablet is a series of accurately produced steps of image density and is readily available at camera and photographic supply stores. The human eye is a very good null detector. This means

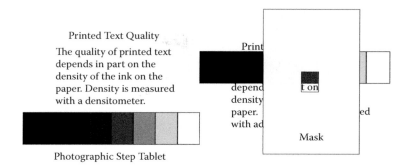

FIGURE 2.7 The visual measurement of print density.

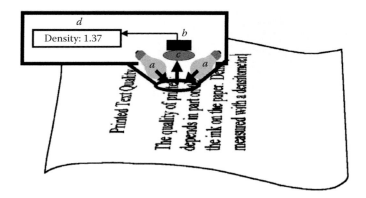

FIGURE 2.8 Densitometer showing light sources (*a*) at 45 degrees, a light detector (*b*) at zero degrees, a filter (*c*), and an output display (*d*).

that even though we are not able to judge the density of an image accurately by simply looking at it, we are quite capable of judging whether or not two images placed side by side are the same or different. Thus, by masking the step tablet and the printed image, as illustrated in Figure 2.7, it is easy to match a gray level of the photographic step tablet with the printed image.

Several companies manufacture instruments for measuring image density. These instruments, called densitometers, range in price from a few hundred to several thousand dollars. Figure 2.8 illustrates a typical densitometer.

Light of brightness I_o is delivered to the paper from light sources typically placed at 45 degrees from the vertical. The electronic light detector detects the reflected light intensity, I, and the electronics of the system combines the values of I and I_o to calculate image density, D, according to Equation 2.1 and Equation 2.2. These equations are the definitions of reflectance, R, and reflection density, D. The density value is then displayed on the output screen.

$$R = I/I_o \qquad (2.1)$$

$$D = -\log(R) \qquad (2.2)$$

The major advantages of a densitometer over visual matching with a step tablet are convenience and rapidity of operation. Both techniques provide comparable accuracy and precision when measuring the density of black and white images. Color image density is much more difficult to measure by the visual matching technique and requires many more reference samples and more practice to provide accurate results. Nevertheless, books and charts are readily available and in extensive use for performing visual color matching, particularly in situations in which an electronic instrument is impractical.[10] Forensic pathologists, biologists, and geologists often measure color by matching to standard color charts.

A color densitometer is not much more complex than a monochrome densitometer. In general, it requires only the addition of red, green, and blue filters. A monochrome densitometer typically incorporates a filter at location (c) in Figure 2.8 to make the instrument respond to the spectrum of light in same way as the average human observer. By adding a red, green, and blue filter, the densitometer can make accurate and rapid measurements of color density as well as black and white density. Inexpensive color densitometers make all of the necessary measurements rapidly and automatically and display four density values as illustrated in Figure 2.9.

The density value labeled V in Figure 2.9 is the visual density used in monochrome measurements and represents overall impression of lightness/darkness. The R, G, and B density values are the values measured through red, green, and blue filters. These values correlate approximately with our visual impression of color. How well they correlate with our impression of color depends on the specific filters selected for the R, G, and B measurements, as described below. Often, however, a densitometer is used to measure the amount of cyan, magenta, and yellow inks printed on paper rather than the color that appears to the eye. The red, green, and blue density values are therefore often displayed as shown on the right in Figure 2.9 because the red density correlates strongly with the

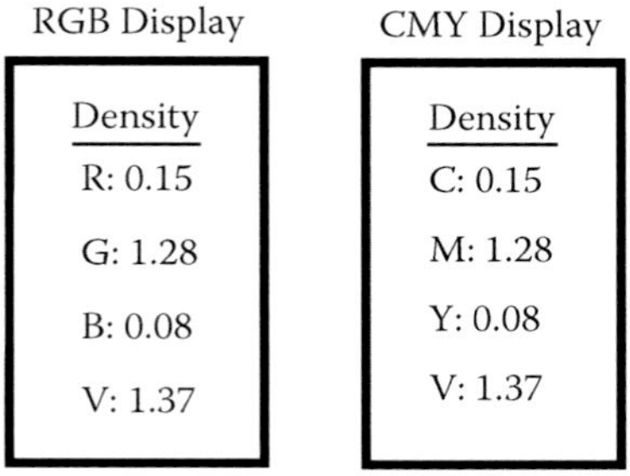

RGB Display	CMY Display
Density	Density
R: 0.15	C: 0.15
G: 1.28	M: 1.28
B: 0.08	Y: 0.08
V: 1.37	V: 1.37

FIGURE 2.9 Output displays often found on color densitometers.

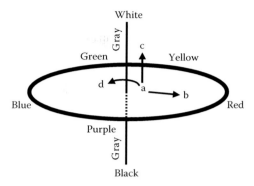

FIGURE 2.10 The Munsell color wheel.

cyan ink, the green density correlates strongly with the magenta ink, and the blue density correlates strongly with the yellow ink.

2.3   COLOR INKS AND COLOR MIXING

Much of the apparent complexity of color printing occurs because of confusion and inconsistencies in the way colors are named. The basic concepts are actually not difficult to understand. Many different schemes are used to name colors, and each is useful in a particular area of application, as described in several introductory texts on color science and technology.[10,11] One of best known schemes for naming colors is the Munsell system, illustrated in Figure 2.10. This system is based on the theory of trichromatic color vision, which says that humans have three kinds of color sensors (cone receptors) in the retina of the eye sensitive to three regions of the spectrum of light. As a result of having three, and only three, kinds of color sensors, we experience all colors in terms of three characteristics. In the Munsell system, these characteristics are called hue, value, and chroma.[10] The hue is what we commonly call color and involves familiar color names such as red, orange, and yellow. The Munsell system uses five color names to describe hue: red, yellow, green, blue, and purple. These five hues can be arranged in a hue circle, and any hue (or "color") that human beings can see falls somewhere in the circle. For example, orange is a hue that falls between red and yellow. Turquoise falls between blue and green.

Color value is analogous to our gray scale in Figure 2.7. Any given hue may be a light hue or a dark hue. A light hue of red, for example, is often called pink. In Figure 2.10, point *a* is a red of medium lightness, *c* is the same red hue but of higher lightness (pink), and *d* is a green hue with the same lightness as *a*.

The third attribute of color is called chroma. This is the degree of purity of the color. The red at point *a* in Figure 2.10 is a low-chroma red such as brick red, but point *b* is a higher-chroma red such as fire engine red. The decrease in purity leads all colors toward the neutral gray scale in the center of Figure 2.10, often called the a-chromatic colors, or colors without hue.

Figure 2.10 and the trichromatic theory of color vision tell us that all colors can be described as if they occupied a location in the three-dimensional space defined by the hue circle, the value axis, and the distance from the axis called the chroma. All colors that human beings are able to see occupy unique locations in this three-dimensional space, and the total volume of space occupied by all visible colors is called the total color gamut of human vision.

2.3.1 PRIMARY COLORS

A consequence of the tri-chromatic theory of human vision is that it is possible to produce all color hues by making mixtures with only three well-selected colors. These well-selected colors are generally called primary colors. This is a somewhat ambiguous term, but it is commonly used by both experts and laypersons. If you ask the average person to name the three primary colors, they often will reply red, blue, and yellow. This is entirely correct within the crayon color scheme of naming colors. However, a much more useful system of color nomenclature is in use by scientists and engineers in which the primary colors are called red, green, and blue. Light sources that produce red, green, and blue light have been found, by trial and error as well as by theory, to have the ability of mixing to make the greatest number of visible colors.[10,11] The higher the chroma of the three light sources, the more colors one can make. Blue, green, and red monochromatic light sources (light of only one wavelength) at 400 nm, 525 nm, and 700 nm, for example, can mix to produce nearly all colors the average observer is able to see. More practical light sources for red, green, and blue are used in television sets and computer monitors. It should be clear that the term *primary color* is only a general concept and not a rigorously defined set of colors.

Printing processes also make use of red, green, and blue primary light to control colors. This is done by starting with the unprinted paper used as the printing substrate. The unprinted paper reflects red, green, and blue light toward our eyes so we see the mixture of the light and call it "white." To control the amount of red light that reflects from the paper, we use an ink that absorbs only the red light. The red light is still our primary color of light, but we control this color of light with an ink that absorbs the red light. When we look at such an ink, the color we do not see is red. This ink looks exactly unlike red, and the traditional name that printers give this hue is cyan, as illustrated in Table 2.1. Similarly, ink that absorbs green light is used to control green light in the printed images. This ink is called magenta. A yellow ink is used to absorb the blue primary of light.

Red, green, and blue are our primary hues of light, sometimes called additive primaries. The corresponding inks (cyan, magenta, and yellow) are often called printer primaries, subtractive primaries, or primary inks. Collectively, these six hues form a hue circle similar to the Munsell hue circle, as illustrated in Figure

TABLE 2.1
The Primary Colors and Primary Inks

The Light Absorbed by the Ink	Name Given to the Hue of Ink Color
Red	Cyan
Green	Magenta
Blue	Yellow

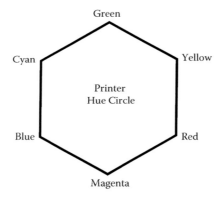

FIGURE 2.11 The descriptive hue circle used by printers, engineers, and scientists. Red, green, and blue are additive primary colors. Cyan, magenta, and yellow are printer primary colors.

2.11. By convention, red is always on the lower right of this hue circle, and green is at the top. Unlike the Munsell system, the printer hue circle is not used as the basis of an exact color naming system. Rather, the printer hue circle is a tool used extensively in qualitative descriptions of color printing. It is a convenient map for orienting oneself in color space, so it is well worth the effort to memorize the printer hue circle. As an example, one can refer to the printer hue circle and say instantly that the color blue is made by printing a combination of cyan and magenta inks, that combining red and green light will make yellow, and that a mix of cyan and yellow ink make green.

It is easy to understand why many people say the three primary colors are red, blue, and yellow. The color called cyan in the printer hue circle looks to most people like a sky blue, the printer blue in common crayon terms is called violet, and magenta is often called a bubble gum shade of red. Therefore, the terms *red*, *blue*, and *yellow* are simply the common crayon names for the printer primary inks called magenta, cyan, and yellow. It is also easy to see why words are inadequate as a means of naming colors. Color measuring instruments that report numbers are much less ambiguous than words.

2.3.2 COLOR DENSITOMETRY

The tri-chromatic nature of human vision provides a rationale for why color densitometers measure density with red, green, and blue filters, as described previously in Figure 2.9. The printer hue circle helps us remember that a red filter is used in a densitometer to measure the cyan ink, as indicated in Figure 2.9. A high red density means that much red light is absorbed, which means the image appears cyan. Red density is useful for measuring the amount of cyan ink on a printed page. The term *cyan density* is often used instead of red density because a high red density appears cyan to the eye. Such terminology is logically inconsistent, but it is in common use. Similarly, green and blue density values are used to measure magenta and yellow inks. Because printers often are concerned with the amount of ink used in printing, they often rely on densitometry as a means of monitoring the ink printed on the page.

Referring to the diagram of the densitometer in Figure 2.8, one uses red, green, and blue (RGB) filters to measure density that correlates well with the amount of cyan, magenta, and yellow (CMY) inks printed on a page. However, if one is more concerned with the visual appearance of the printed color than with the amount of ink used, then a special set of filters must be used instead of the RGB filters. These special filters adjust the spectral response of the instrument so it correlates exactly with the spectral responses of the three cone sensors in the average human eye. The instrument then translates the density measurements into a special color space that correlates well with the visual appearance of the printed image. Such an instrument is called a colorimeter instead of a densitometer.

2.3.3 COLORIMETRY COORDINATES

A colorimeter looks just like an RGB densitometer. However, the colorimeter has special filters so that instead of measuring three density values (red, green, and blue), it measures three numbers called color coordinates [X,Y,Z]. These color coordinates correlate with our visual perception of colors. No two colors will have the same [X,Y,Z] values, and two colors that look identical will have identical [X,Y,Z] values. This is not the case with RGB densitometers. If one is concerned with the behavior of inks on paper, a densitometer is the preferred instrument, but if one is concerned with the measurement of the visual quality of the printed colors, a colorimeter is the preferred instrument. Because the two instruments use two different kinds of filters, it is not possible to convert the results of one instrument into the other. RGB density values cannot be converted to exact [X,Y,Z] values, and [X,Y,Z] values cannot be converted to exact RGB density values. An approximate correlation can be made between the two, but intrinsic mathematical ambiguities limit the accuracy of such a correlation.

The three numbers produced as output from a colorimeter map uniquely and unambiguously onto our perception of color. However, there are several color coordinate systems that do this, and several are in common use. The [X,Y,Z] numbers are commonly used by most commercial colorimeters. These three numbers are defined according to the integral functions shown below. The term $P(\lambda)$ is the spectral energy distribution of the light under which the printed image is viewed, and $R(\lambda)$ is the spectral reflectance of the printed image.

$$100 = \int_{400}^{700} P(\gamma) \cdot \bar{y}(\gamma) d\gamma \qquad (2.3)$$

$$X = K \cdot \int_{400}^{700} R(\gamma) \cdot P(\gamma) \cdot \bar{x}(\gamma) d\gamma \qquad (2.4)$$

$$Y = K \cdot \int_{400}^{700} R(\gamma) \cdot P(\gamma) \cdot \bar{y}(\gamma) d\gamma \qquad (2.5)$$

$$Z = K \cdot \int_{400}^{700} R(\gamma) \cdot P(\gamma) \cdot \bar{z}(\gamma) d\gamma \qquad (2.6)$$

The terms $\bar{x}(\gamma), \bar{y}(\gamma), \bar{z}(\gamma)$ are called color-matching functions and are functions that are linearly related to the spectral response of the three color sensors in the human eye.

With the correct filters, the colorimetric instrument automatically performs the calculations indicated in Equation 2.3 through Equation 2.6, and one only has to note the X, Y, and Z color values in just the same way one would note the RGB density values when using an ordinary densitometer. The [X,Y,Z] values are more like reflectance values, R in Equation 2.1, than density values. A high value of Y, for example, means a low green density and a high reflectance of green light. Indeed, a densitometer that measures the visual density, V in Figure 2.9, actually measures Y and calculates $V = -\log(Y)$.

Most colorimeters can present the color coordinates [X,Y,Z] and also other useful color coordinates. Many commercial instruments convert the [X,Y,Z] coordinates into [L*,a*,b*] coordinates, which are more useful for describing the color difference we see between two printed colors. The reader is referred to texts on colorimetry for more information about color theory and colorimetry.[10,11]

2.4 THE PRINTING OF PICTORIAL IMAGES

The quality of a pictorial image depends on the quality of the reproduction of gray and color tones. The reproduction of gray and color tones is influenced by the same printing factors that influence text printing. However, the optimum values of these factors for text printing and pictorial printing are generally not the same. Factors that produce high-resolution, high-density letters on clean, white paper often lead to pictorial images of poor tone and color quality.[12,13]

Most printing technologies are inherently incapable of printing gray tones. They do well when printing a text letter at full density or when printing text at all. The 18th-century printing process used by Benjamin Franklin, for example, employed a relief printing plate with raised letters that pressed ink into paper. Text quality was good, but no gray tones could be produced. This was the case for all printing processes until the development, during the 19th century, of the photographic process for printing film negatives onto photographic print paper. The photographic process is intrinsically able to print a continuous range of gray and is thus often referred to as a continuous tone process. Mechanical printing processes, including those used for desktop digital printing, are intrinsically bi-level processes. They either deliver ink at a selected location or they do not. With few exceptions, they are unable to print continuous tones.

To simulate the appearance of gray tones, printers have long used a variety of techniques for fine-scale distribution of ink, as illustrated in Figure 2.12. Many different techniques for simulating tone have been developed over the past several centuries.[14] The most common technique currently in use for both commercial and desktop printing is the halftone technique, illustrated in Figure 2.13.

By converting a continuous tone image into a halftone image, as illustrated in Figure 2.13, the printing process does what it intrinsically does best. It prints solid ink on white paper, simulating gray tone by the size or closeness of the printed dots. It would seem that the ability to print crisp, well-formed dots, would depend on the same factors as those for printing text. Indeed, if the halftone dots are as large as in the illustration of Figure 2.13, this is so. However, high-quality halftone images require the printing of very small halftone dots. The continuous tone illustration in Figure 2.13(A) is actually not a continuous

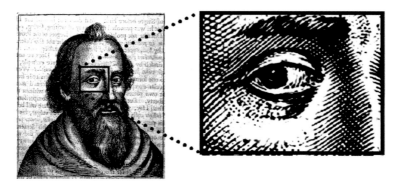

FIGURE 2.12 A print by William Downing, printer of Juvenal (tr. to B. Holyday) Decimus Junius Juvenalis, (Oxford, 1673). (Used with permission from the Melbert B. Cary, Jr. Graphic Arts Collection, Rochester Institute of Technology.)

Continuous Tone Photograph   Halftone Copy

(a)   (b)

FIGURE 2.13 Illustration of a continuous tone image and a corresponding halftone image. (Courtesy of Frank Cost.)

tone photograph at all. It is a halftone printed at 150 halftone dots per linear inch. This can be seen by examining Figure 2.13(A) with a strong magnifying glass (at least 6× recommended). It is clear that high quality halftone images require that dots be printed well and also that they be printed much smaller than the text characters used in document printing. The need to print very small dots makes the requirements for halftone pictorial quality different from the requirements for text quality.

2.4.1 Halftone Printing

A halftone image can be indistinguishable from a continuous tone photograph if the halftone dots are smaller than the eye can resolve. In this case, the eye perceives only the average reflectance of light from the image, as illustrated in Figure 2.14. Each ray of light from the continuous tone image is equally bright, but the halftone reflects rays of two types. Some of the rays are reflected from the paper and are very bright. Some of the rays are reflected from the ink dots and are very dim. If the dots are small enough, the observer sees only the average brightness from the combined ink and paper rays of reflected light.

Quantitatively, the lightness of the image can be described in terms of its reflectance, which was defined in Equation 2.1 as the ratio of the reflected light to the incident light, $R = I/I_o$. The continuous tone image on the left side of Figure 2.14 has a single reflectance value, R. The halftone image on the right has an average reflectance, R, given by the weighted average of the ink and paper reflectance, as illustrated in Equation 2.7, where F is the fraction of the area of

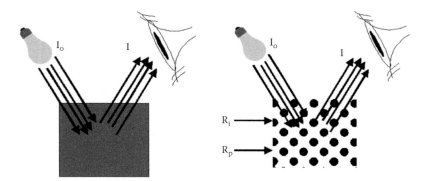

FIGURE 2.14 Viewing a continuous tone photograph and a halftone print of the same average reflectance.

the image that is covered by dots and $(1 - F)$ is the area of the bare paper between the dots.

$$R = F \cdot R_i + (1 - F) \cdot R_p \tag{2.7}$$

The human visual system responds approximately logarithmically to changes in reflectance. Thus, density, which is the negative log of reflectance (see Equation 2.2), is often preferred when evaluating the appearance of a printed gray scale. In terms of density, F can be solved from Equation 2.7 as shown in Equation 2.8.

$$F = \frac{10^{-D_p} - 10^{-D}}{10^{-D_p} - 10^{-D_i}} \tag{2.8}$$

where D_i is the density of the ink, D_p is the density of the paper, and D is the average density of the image.

Densitometer readings are often calibrated in such a way that the paper is assigned the relative density value of 0.00. In that case, Equation 2.9 is the relationship between the density of the image and the area fraction of the image that is covered by dots.

$$F = \frac{1 - 10^{-D}}{1 - 10^{-D_i}} \tag{2.9}$$

Equation 2.7 through Equation 2.9 are each variations of what is often called the Murray–Davies equation.[11,13] The Murray–Davies equation shows how one can control a printing device to produce any desired gray level by controlling the dot area fraction, F. One can calibrate the printing system by measuring the ink and paper reflectance values and then control the output gray level by controlling the dot area fraction, F, sent to the printer. In practice, when a user selects the desired gray level, R, the software that controls the printer translates the value of R into the dot area command, F, that is sent to the printer. This software process is called raster image processing, or RIP, and we often refer to the process as ripping the image.[15]

The Murray–Davies equation is a very simple description of halftone imaging, and, not surprisingly, it is a good description only for ideal halftones. Real halftones suffer from a variety of physical and optical effects that render Equation 2.7 through Equation 2.9 only poor approximations of printed images. Failure of the Murray–Davies equation, regardless of cause, is often called dot gain.[12,13,15] This is described in more detail in Section 2.5.

2.4.2 COLOR HALFTONE PRINTING

The three primary printing inks cyan, magenta, and yellow are used to print color images. Mixtures of these three inks can produce any desired hue over a wide range of shades. There are two ways one can mix inks in a printed image, and they produce significantly different results. Color photography and color photographic printing involve the physical mixing of cyan, magenta, and yellow dyes within a single gelatin layer. The mechanical processes of printing ink on paper involves printing three overlapping halftone patterns, one for each of the primary inks. This is true for the processes of printing as well as for digital desktop printing. The way in which the separate dot patterns for cyan, magenta, and yellow are combined on the printed paper has a significant effect on the quality of the printed image. One major phenomenon that must be considered is the moiré effect.

The Moiré Effect: When more than one halftone pattern is printed on a paper, the way the two patterns are arranged has a major impact on color and tone reproduction. For example, suppose the halftone pattern shown on the left side of Figure 2.15 is to be printed along with a second, identical pattern. If we print the second pattern so that the second set of dots falls exactly between the first set, we have the much darker image on the right side of Figure 2.15. However, if the second set of dots is printed in perfect registration with the first set, the result is indistinguishable from the original pattern on the left. It is clear from this illustration that the location of the dots must be controlled appropriately to control tone and color printing.

FIGURE 2.15 Dots on dots and dots beside dots with perfectly black ink.

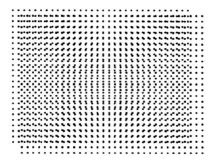

FIGURE 2.16 Illustration of moiré caused by superimposing two halftone patterns that differ by 2% in their LPI.

The example in Figure 2.15 illustrates the extreme difference in appearance between printing dots-on-dots and dots-beside-dots. To do either, the registration of the dot patterns must be in perfect control. Precision dot placement of this kind is not achievable in most printing systems for a variety of reasons, including the intrinsic elasticity of paper itself. For example, if the paper stretches a small amount between printing the first and second dot pattern, a very severe non-uniformity can result, as illustrated in Figure 2.16. This effect is called moiré. With 150-lines-per-inch (LPI) halftones, this effect can be very severe and objectionable, with less than a percent difference in the halftone LPI when attempting to register two halftone patterns exactly. For this reason, it is generally impractical to attempt to print either the dot-on-dot or the dot-between-dot patterns.

Another example of moiré is observed when two identical dot patterns are superimposed with a rotation angle between them, as illustrated in Figure 2.17.

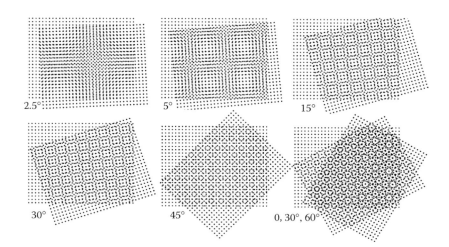

FIGURE 2.17 The moiré effect and screen rotation.

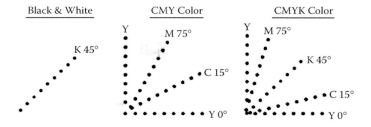

FIGURE 2.18 Commonly used angles for halftone printing.

The rotation angle is called the *screen angle*, a term originating from the old photographic process of using a halftone screen to convert photographic images into halftone images.[13-15] The greater the screen angle, the smaller the moiré pattern is; thus, to minimize the moiré effect, halftone patterns are superimposed with a large screen angle.

Three patterns are required to print color images, so a typical technique is to print the yellow at zero degrees, the cyan at 30 degrees, and the magenta at 60 degrees, as illustrated in Figure 2.18. The resulting pattern of rosettes, illustrated in the lower right configuration of Figure 2.17, is too small for the average observer to see for 150 LPI halftone patterns at a typical reading distance of 35 cm. If black ink is printed in addition to the three primary inks, the typical scheme places yellow at zero degrees, cyan at 15 degrees, black to 45 degrees (30 degrees from cyan), and magenta at 75 degrees (30 degrees from black and 15 degrees from yellow). Using 15 degrees between yellow and cyan and between yellow and magenta would be expected to lead to undesirable moiré, but the human eye is less sensitive to yellow patterns or moiré involving yellow.[12]

2.4.3 HALFTONE COLOR CALCULATIONS

As illustrated in Figure 2.15 and Figure 2.16, the tone and color reproduced by superimposing halftone patterns depend strongly on the way they are superimposed and on the quality of the registration between the patterns. Because exact registration is not a practical goal in color printing, the alternative for color reproduction is to print the halftone dots in random registration. This random dot placement is approximated by the same rotated screen technique used to avoid moiré. As shown in Figure 2.19, using 0 degrees, 30 degrees, and 60 degrees for yellow, cyan, and magenta dots, all degrees of dot overlap occur. This quasi-random pattern of overlap is the key to accurate color calculations with halftones. To calculate the total integrated color one sees when viewing a halftone image at a normal reading distance, we need to add up all the microcolors in the microregion illustrated in Figure 2.19. This is a three-step process first described by Neugebauer in 1937.[16]

Step I: Calculate the Color Area Fractions. Three signals (F_c, F_m, and F_y) are formed in the color RIP process and are sent to the printer. These three signals are the fractional dot area (F) values in Equation 2.7 applied to each of the three primary

Green Overlap Color   Black Overlap Color

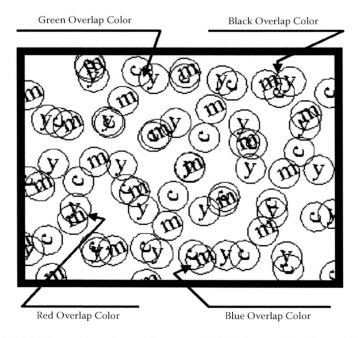

Red Overlap Color   Blue Overlap Color

FIGURE 2.19 The quasi-random registration of CMY dots at 30°, 60°, and 0° screen rotations. Note the regions of RGB and black overlap.

inks: cyan, magenta, and yellow. This causes the printing device to print three halftone patterns: a pattern for cyan at dot area fraction F_c, a magenta pattern at F_m, and a yellow pattern at F_y. The random overlap of these three patterns results in eight color area fractions, shown in Table 2.2. A region with no cyan $(1 - F_c)$, no magenta $(1 - F_m)$, and no yellow $(1 - F_y)$ is white. The fraction of the image that is white is the product $f_w = (1 - F_c)(1 - F_m)(1 - F_y)$. The area fractions of the other seven colors are similarly calculated, as shown in Table 2.2. These eight colors are called the Neugebauer primary colors.

Step II: Measure the eight Neugebauer primary colors. We print test patches of each of the eight Neugebauer primary colors. We can measure the reflection spectrum of each primary sample, $R_w(\lambda)$, $R_c(\lambda)$, $R_m(\lambda)$, $R_y(\lambda)$, $R_r(\lambda)$, $R_g(\lambda)$, $R_b(\lambda)$, and $R_k(\lambda)$. We also can measure the color coordinates for each sample,

$$[X_w, Y_w, Z_w], [X_c, Y_c, Z_c].... [X_k, Y_k, Z_k].$$

Step III: Calculate the Printed Color. We apply an expanded version of the Murray–Davies Equation 2.7 to calculate the spectrum of the printed image, $R(\lambda)$ as shown in Equation 2.10. Equation 2.11 is a shorthand way to represent this calculation, in which the subscript 1 through subscript 8 stand for colors w through k.

TABLE 2.2
The Eight Neugebauer Colors Produced by Printing Random or Pseudo-Random Dots of Cyan, Magenta, and Yellow Ink. The Three Area Fractions of Ink (F_c, F_m, and F_y) Produce the Color Area Fractions (f_w, f_c, f_m, f_y, f_r, f_g, f_b, f_k)

Color	Area Fraction
White	$f_w = (1 - F_c)(1 - F_m)(1 - F_y)$
Cyan	$f_c = F_c(1 - F_m)(1 - F_y)$
Magenta	$f_m = (1 - F_c) \, F_m(1 - F_y)$
Yellow	$f_y = (1 - F_c)(1 - F_m) \, F_y$
Red	$f_r = (1 - F_c) \, F_m \, F_y$
Green	$f_g = F_c(1 - F_m) \, F_y$
Blue	$f_b = F_c(1 - F_m) \, F_y$
Black	$f_k = F_c \, F_m \, F_y$

$$R(\gamma) = f_w \cdot R_w(\gamma) + f_c \cdot R_c(\gamma) + f_m \cdot R_m(\gamma) + f_y \cdot R_y(\gamma) +$$
$$f_r \cdot R_r(\gamma) + f_g \cdot R_g(\gamma) + f_b \cdot R_b(\gamma) + f_k \cdot R_k(\gamma) \tag{2.10}$$

$$R(\gamma) = \sum_{i=1}^{8} f_i R_i(\gamma) \tag{2.11}$$

Color Equation 2.4 through Equation 2.6 can be applied to the reflection spectrum, R(y), to calculate the color coordinates of the image, [X,Y,Z]. Alternatively, the measured color coordinates of the eight Neugebauer primary samples can be combined to find the coordinates of the printed image, as shown in Equation 2.12 through Equation 2.14.

$$X = \sum_{i=1}^{8} f_i X_i \tag{2.12}$$

$$Y = \sum_{i=1}^{8} f_i Y_i \tag{2.13}$$

$$Z = \sum_{i=1}^{8} f_i Z_i \qquad\qquad (2.14)$$

Equation 2.10 through Equation 2.14 are collectively known as the Neugebauer equations. They are the color equivalent of the Murray–Davies Equation 2.7.

The practical application of this kind of calculation involves selecting the color one wishes to print, [X,Y,Z], and inverting the calculations shown above to determine the three ink area fractions, F_c, F_m, and F_y, that must be sent to the printer. The simple Murray–Davies equation is easily inverted to calculate F given R, R_p, and R_i. However, the Neugebauer equations are not straightforward to invert. One must rely on numerical and statistical techniques to do this. One straightforward method is to define a wide range of values for F_c, F_m, and F_y and calculate the color coordinate [X,Y,Z] for each combination. The result is then analyzed statistically to construct a set of empirical equations for predicting F_c, F_m, and F_y from [X,Y,Z].

2.5   TONE AND COLOR CONTROL

The Murray–Davies equation and the Neugebauer equations are idealized descriptions of perfectly well-behaved halftones. However, real printed images almost never behave exactly as described by Murray–Davies and Neugebauer, and some amount of correction is needed to achieve acceptable image quality. The most common problem encountered is the dot-gain phenomenon.[12,13,15]

2.5.1   THE DOT-GAIN PHENOMENON

When printing a halftone image with a single ink, the actual printed image is usually darker than predicted by the Murray–Davies equation. This is illustrated in Figure 2.20, where F_n is the dot area command sent to the digital printer. The printer responds by printing halftone dots, but the measured reflectance of the printed image (Figure 2.20b) is darker than expected. The phenomenon is not unique to digital printing. It was originally observed by commercial printers more than a century ago, before much was known about the underlying physical and optical processes involved. Early printers assumed the image was printed too dark because the halftone dots spread out in the printing process. If this occurred, the size of the printed dots, F_a, must be larger than the nominal dot size, F_n, that was intended. Thus, the darkening phenomenon illustrated in Figure 2.20 was called the dot-gain effect. This term is still in use, even though the underlying causes are more involved than a simple increase in the physical size of the printed dot.

If one assumes the Murray–Davies equation would be correct if one used the correct dot size in Equation 2.7, then one can rearrange Equation 2.7 and solve for the apparent dot size, F_a, as shown in Equation 2.15. Thus, by

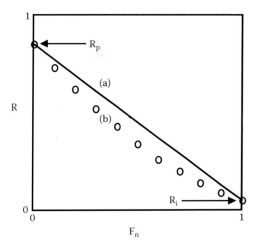

FIGURE 2.20 The dot-gain phenomenon. For a series of nominal dot area fractions, F_n, sent to a printer, line (a) is the result predicted by the Murray–Davies equation and data points (b) are the actual printed image.

measuring the actual printed reflectance, R, the reflectance of the paper, R_p, and the reflectance of the solid ink, R_i, one may calculate the apparent size of the halftone dot, F_a.

$$F_a = \frac{R_p - R}{R_p - R_i} \tag{2.15}$$

This leads to the quantitative definition of dot gain given in Equation 2.16. Dot gain, defined in this way, is a useful metric for characterizing and calibrating the printing process, and knowledge of dot gain is essential to control the printing process for maximum pictorial image quality.

$$DG = F_a - F_n \tag{2.16}$$

2.5.2 DOT-GAIN MECHANISMS

There are two primary causes of the dot-gain phenomenon. The physical spread of ink on paper is one cause. In general, the decreased reflection density that results when the ink spreads out on the paper is more than offset by the increased area coverage, F, and the overall image is darker.

The second major cause of dot gain is an optical effect. As illustrated in Figure 2.21, light used to view the halftone image can enter the paper by passing through a halftone dot or by entering between the dots. The light that enters the paper at (B) in Figure 2.21 scatters around in the paper and eventually emerges

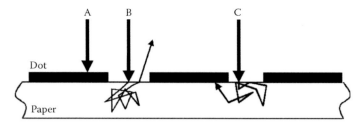

FIGURE 2.21 Illustration of optical dot gain.

as reflected light between the dots. Some fraction of the light, however, enters as shown in (C) and scatters under a dot and is absorbed. This increases the fraction of light that is absorbed by the halftone dots and results in a darker image. Thus, the effect is similar in appearance to a physical dot gain and is often called an optical dot gain. It is also called the Yule–Nielsen effect.[17]

Much research has been devoted to both physical and optical dot gain and to the characteristics of the paper, the ink, and the dot patterns that influence it. In general, most common non-coated papers show about the same amount of dot-gain, both physical and optical. Coated papers show less of an effect, and special substrates designed to minimize these and other artifacts of the printing process show much less dot gain. Inkjet, in particular, benefits significantly from the use of special substrates. Nevertheless, some amount of dot gain is always observed, and systems in common use for digital desktop printing have been well optimized to compensate for these effects.

2.5.3 THE YULE–NIELSEN CORRECTION

There are several ways to calibrate a printer to correct for the dot-gain phenomenon. The most general approach is empirical. One prints a set of known samples with a set of known print commands, F_n, and measures the resulting output reflectance, R. Then a statistical polynomial regression can be used to fit the data. The resulting polynomial is called a printer model, or sometimes a look-up-table (LUT), and provides the necessary tone correction in the printing process. The process that does this is part of what is called "color management."[18]

When multiple colors are printed, the polynomial regression process can become quite complicated. Before the advent of powerful desktop computers, such complexity required a much simpler solution. In the early 1950s a simple, empirical equation was suggested as a correction for dot gain. This is called the Yule–Nielsen equation, shown below as Equation 2.17.[13]

$$R = \left[F \cdot R_i^{1/n} + (1 - F) \cdot R_p^{1/n} \right]^n \qquad (2.17)$$

The Yule–Nielsen equation is a simple polynomial that looks much like the Murray–Davies equation but with a single power factor called the Yule–Nielsen

TABLE 2.3
Halftone Data for a 60-LPI Halftone Printed with a 600-dpi Printer.
F_n = **Nominal Dot Fraction Command. R = Reflectance of Printed**
Sample

F_n	0	0.1	0.2	0.3	0.4	0.5	0.6	0.7	0.8	0.9	1.0
R	0.8	0.7	0.6	0.5	0.4	0.3	0.2	0.2	0.1	0.0	0.0
	5	3	2	2	3	4	6	0	4	9	5

n-factor. By fitting this equation to measured data, the value of n is determined. The n value is then a single calibration constant for the printing process. The dot gain, DG, defined in Equation 2.16 varies with the gray level of the printed image. However, the Yule–Nielsen n factor is useful as a single index of dot gain for the printing process at any gray value. A value of n = 1 means zero dot gain, and any n > 1 means non-zero dot gain. The Yule–Nielsen equation, and variations of it, are often still used to characterize both commercial and desktop printers.

Using the Yule–Nielsen n factor is illustrated by the data in Table 2.3. These data are the experimental data shown in Figure 2.20. The halftone dots were printed using an inkjet 600-dpi printer to form the halftone dots at 60 halftone dots per inch. The printer was sent 11 nominal dot fractions (F_n in Table 2.3), and the printer responded by printing 11 gray patches. The reflectance, R, of each patch is also shown in Table 2.3.

The value of the Yule–Nielsen n factor used in Equation 2.17 can be found experimentally by a spreadsheet calculation using the data in Table 2.3. The first step is to note the values of R_i and R_p. The value of R_p is shown in Table 2.3 as the value of R at $F_n = 0$. Thus, R_p = 0.85. Similarly, the solid ink reflectance is R_i = 0.05. The values of n typically observed for most halftone images is in the range $1 < n < 5$. In the spreadsheet, we can define a trial range of n values we believe will include the correct value. Later, if the calculation fails to identify the correct value of n, we can adjust the range. In this example, the trial range of n values was chosen from 1.70 to 1.90, as shown in Table 2.3. The nominal values of the dot area fraction, F_n, range from 0 to 1, as shown in the table. For each combination of n and F_n, Equation 2.17 was used to calculate the reflectance, R.

The values of R shown in Table 2.3 were compared with the values of R measured experimentally. For each value of n and F_n, the deviation, δ, between the calculated and experimental value of R was determined. The squared deviation values, δ^2, are shown in Table 2.4.

For each value of n, the values of δ^2 were averaged over the range of F_n values. The standard deviation, σ, is the square root of the average value of δ^2. These values are shown on the right of Table 2.5. Figure 2.22 shows the value of σ vs. the trial value of n. The standard deviation reaches a minimum at n = 1.82. This is our experimental estimate of the n value characteristic of this printing process.

TABLE 2.4
Values of R Calculated with Equation 2.17 Using $R_p = 0.85$ and $R_i = 0.05$. Experimentally Measured Values of R for Each Value of F_n are Shown as the Last Row of the Table

n/F_n	0.0	0.1	0.2	0.3	0.4	0.5	0.6	0.7	0.8	0.9	1.0
1.70	0.85	0.77	0.69	0.61	0.53	0.45	0.37	0.29	0.21	0.13	0.05
1.72	0.85	0.735	0.628	0.528	0.435	0.349	0.272	0.203	0.143	0.091	0.05
1.74	0.85	0.735	0.627	0.526	0.433	0.348	0.27	0.202	0.142	0.091	0.05
1.76	0.85	0.734	0.625	0.524	0.431	0.346	0.269	0.2	0.141	0.09	0.05
1.78	0.85	0.733	0.624	0.523	0.429	0.344	0.267	0.199	0.14	0.09	0.05
1.80	0.85	0.733	0.623	0.521	0.428	0.343	0.266	0.198	0.139	0.09	0.05
1.82	0.85	0.732	0.622	0.52	0.426	0.341	0.264	0.197	0.138	0.089	0.05
1.84	0.85	0.731	0.621	0.518	0.425	0.339	0.263	0.196	0.137	0.089	0.05
1.86	0.85	0.731	0.619	0.517	0.423	0.338	0.262	0.195	0.137	0.088	0.05
1.88	0.85	0.73	0.618	0.516	0.422	0.336	0.26	0.193	0.136	0.088	0.05
1.90	0.85	0.729	0.617	0.514	0.42	0.335	0.259	0.192	0.135	0.088	0.05
R =	**0.85**	**0.73**	**0.62**	**0.52**	**0.43**	**0.34**	**0.26**	**0.20**	**0.14**	**0.09**	**0.05**

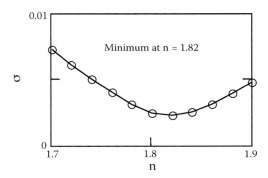

FIGURE 2.22 Standard deviation in reflectance vs. trial value on n from the data in Table 2.1 through Table 2.4.

TABLE 2.5
The Values of Squared Deviation, δ^2, for Each Combination of F_n and n

n/F_n	0.0	0.1	0.2	0.3	0.4	0.5	0.6	0.7	0.8	0.9	1.0	σ
1.70	0	$1.6 \cdot 10^{-3}$	$4.9 \cdot 10^{-3}$	$8.1 \cdot 10^{-3}$	0.01	0.012	0.012	$8.1 \cdot 10^{-3}$	$4.9 \cdot 10^{-3}$	$1.6 \cdot 10^{-3}$	0	76.0×10^{-3}
1.72	0	$2.931 \cdot 10^{-5}$	$6.17 \cdot 10^{-5}$	$5.636 \cdot 10^{-5}$	$2.097 \cdot 10^{-5}$	$8.665 \cdot 10^{-5}$	$1.434 \cdot 10^{-4}$	$8.441 \cdot 10^{-6}$	$6.305 \cdot 10^{-6}$	$1.713 \cdot 10^{-6}$	0	6.14×10^{-3}
1.74	0	$2.198 \cdot 10^{-5}$	$4.361 \cdot 10^{-5}$	$3.505 \cdot 10^{-5}$	$8.04 \cdot 10^{-6}$	$5.736 \cdot 10^{-5}$	$1.081 \cdot 10^{-4}$	$2.611 \cdot 10^{-6}$	$2.587 \cdot 10^{-6}$	$7.314 \cdot 10^{-7}$	0	5.05×10^{-3}
1.76	0	$1.578 \cdot 10^{-5}$	$2.885 \cdot 10^{-5}$	$1.902 \cdot 10^{-5}$	$1.267 \cdot 10^{-6}$	$3.454 \cdot 10^{-5}$	$7.848 \cdot 10^{-5}$	$1.3 \cdot 10^{-7}$	$5.368 \cdot 10^{-7}$	$1.737 \cdot 10^{-7}$	0	4.03×10^{-3}
1.78	0	$1.067 \cdot 10^{-5}$	$1.729 \cdot 10^{-5}$	$7.999 \cdot 10^{-6}$	$3.039 \cdot 10^{-7}$	$1.778 \cdot 10^{-5}$	$5.413 \cdot 10^{-5}$	$7.417 \cdot 10^{-7}$	$1.374 \cdot 10^{-8}$	$5.528 \cdot 10^{-11}$	0	3.15×10^{-3}
1.80	0	$6.61 \cdot 10^{-6}$	$8.786 \cdot 10^{-6}$	$1.748 \cdot 10^{-6}$	$4.822 \cdot 10^{-6}$	$6.714 \cdot 10^{-6}$	$3.471 \cdot 10^{-5}$	$4.207 \cdot 10^{-6}$	$8.882 \cdot 10^{-7}$	$1.747 \cdot 10^{-7}$	0	2.50×10^{-3}
1.82	0	$3.551 \cdot 10^{-6}$	$3.198 \cdot 10^{-6}$	$2.51 \cdot 10^{-8}$	$1.451 \cdot 10^{-5}$	$1 \cdot 10^{-6}$	$1.989 \cdot 10^{-5}$	$1.03 \cdot 10^{-5}$	$3.041 \cdot 10^{-6}$	$6.65 \cdot 10^{-7}$	0	2.26×10^{-3}
1.84	0	$1.458 \cdot 10^{-6}$	$3.971 \cdot 10^{-7}$	$2.604 \cdot 10^{-6}$	$2.908 \cdot 10^{-5}$	$3.105 \cdot 10^{-7}$	$9.372 \cdot 10^{-6}$	$1.883 \cdot 10^{-5}$	$6.364 \cdot 10^{-6}$	$1.441 \cdot 10^{-6}$	0	2.52×10^{-3}
1.86	0	$2.911 \cdot 10^{-7}$	$2.604 \cdot 10^{-7}$	$9.268 \cdot 10^{-6}$	$4.824 \cdot 10^{-5}$	$4.336 \cdot 10^{-5}$	$2.874 \cdot 10^{-6}$	$2.958 \cdot 10^{-5}$	$1.075 \cdot 10^{-5}$	$2.476 \cdot 10^{-6}$	0	3.14×10^{-3}
1.88	0	$1.425 \cdot 10^{-8}$	$2.668 \cdot 10^{-6}$	$1.981 \cdot 10^{-5}$	$7.173 \cdot 10^{-5}$	$1.279 \cdot 10^{-5}$	$1.299 \cdot 10^{-7}$	$4.24 \cdot 10^{-5}$	$1.612 \cdot 10^{-5}$	$3.746 \cdot 10^{-6}$	0	3.92×10^{-3}
1.90	0	$5.919 \cdot 10^{-7}$	$7.507 \cdot 10^{-6}$	$3.403 \cdot 10^{-5}$	$9.931 \cdot 10^{-5}$	$2.539 \cdot 10^{-5}$	$8.917 \cdot 10^{-7}$	$5.711 \cdot 10^{-5}$	$2.238 \cdot 10^{-5}$	$5.227 \cdot 10^{-6}$	0	4.79×10^{-3}

This technique for determining the value of n is simple and can be performed with a spreadsheet. Once the value of n is found, it is easy to calculate F_n. The dot area fraction, F_n, one must send to the printer to print any chosen value of R is easily calculated using the rearranged form of Equation 2.17. For example, if we want to print a reflectance of R = 0.234, we would use Equation 2.18 to determine that we should send the command F_n= 0.643 to the printer.

$$F_n = \frac{R_p^{1/n} - R^{1/n}}{R_p^{1/n} - R_i^{1/n}} \tag{2.18}$$

2.5.4 YULE–NIELSEN COLOR CALIBRATION

Dot gain also occurs in color printing, and the Yule–Nielsen n-factor can be applied to the Neugebauer color Equation 2.11 through Equation 2.14, as shown in Equation 2.19 through Equation 2.22 below. One may choose to apply the n-factor to the reflection spectra for the eight Neugebauer primary colors, as illustrated in Equation 2.19, or one may apply the factor to the individual colorimetric numbers, as shown in Equation 2.20 through Equation 2.22.

$$R(\gamma) = \left[\sum_{i=1}^{8} f_i R_i^{1/n}(\gamma) \right]^n \tag{2.19}$$

$$X = \left[\sum_{i=1}^{8} f_i X_i^{1/n} \right]^n \tag{2.20}$$

$$Y = \left[\sum_{i=1}^{8} f_i Y_i^{1/n} \right]^n \tag{2.21}$$

$$Z = \left[\sum_{i=1}^{8} f_i Z_i^{1/n} \right]^n \tag{2.22}$$

It is important to remember that the n-factor is an empirical correction and is not based on physical theory. If the color coordinates [X,Y,Z] are calculated from the reflection spectrum R(λ) (see Section 2.4.3), the results will not be the same as the [X,Y,Z] values found by applying the n-factor in Equation 2.20 through Equation 2.22. One must determine experimentally which correction technique works best for any given printing system.

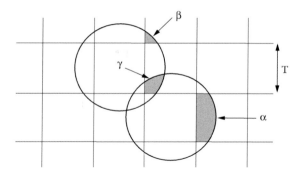

FIGURE 2.23 Geometric model of ink spread. Adapted from Pappas and Pappas, Dong, and Neuhoff.[19,20]

In general, the *n*-factor is a useful tool for characterizing a printer. However, it is not always used as a means of calibrating a printer and managing practical color reproduction. Instead, more extensive experiment data are collected for the printer over a wide range of colors, and an empirical look-up table is developed through advanced statistical techniques of color management.[18]

2.5.5 MECHANISTIC MODELS AND DOT GAIN

Two types of models are used to describe halftone printing. One type of model, the empirical model, is based on statistical analysis of a large set of printed samples. These empirical models are useful for the practical calibration of a printer and for developing methods for color management. The other type of model is the mechanistic model. A mechanistic model attempts to describe the underlying physical or optical mechanisms of the printing process. Mechanistic models are often used to guide scientists and engineers in the design of improved printing devices. However, mechanistic models generally tend to be more complex than empirical models and therefore are not often used for printer calibration or color management. Many examples of mechanistic models have been published. Some describe the way inks penetrate and spread in paper. Others describe the scattering of toner. Numerous models have been published to describe the optical scattering of light in paper. Three examples of mechanistic printer models are presented below.

Geometric Model of Physical Dot Gain: An example of a simple geometric model of physical dot gain is shown in Figure 2.23.[20,21] This model compares the idealized matrix of printer locations to the circular shape of ink dots. For such an idealized process, the geometry of physical dot gain can be solved exactly. The printer is able to address the center of each square of dimension T, but the dot that is printed is of radius r > T. This leads to exactly three geometric shapes, α, β, and γ, each with an area that can be calculated by knowing only T and r. These three types of regions can be used to add up all of the dot gain and dot overlap in the idealized printing process. Such models can provide a very accurate

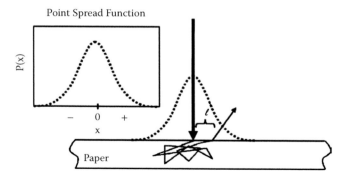

FIGURE 2.24 Illustration of light scattering and the point spread function.

calibration for physical dot gain for a printing process that behaves exactly as described by the model. However, this kind of model is seldom a practical improvement over empirical statistical techniques of printer calibration because most real printers do not behave exactly in the way described by the model. This kind of model is much more useful as a guide in product development than it is for applied color management.

Convolution Model of Optical Dot Gain: Mechanistic models of optical dot gain, illustrated in Figure 2.21, have been explored extensively in recent years.[17,22] The spreading of light can be characterized by a point spread function, illustrated in Figure 2.24. The point spread function is a description of the average distance, ℓ, that light scatters in the paper before it returns to the surface as reflected light. Some light travels a long distance and some a short distance. The bell-shaped curve illustrated in Figure 2.24 is a summary of the distances traveled by all the light entering at location $x = 0$. This bell-shaped curve is called the point spread function for light scattering in the paper and is represented mathematically as $P(x)$.

The point spread function is used to calculate the magnitude of optical dot-gain by combining it with a function that describes the distribution of halftone dot pattern, $T(x)$, illustrated in Figure 2.25. This function describes the fraction of the light that is transmitted through an ink dot, T, as a function of the dot

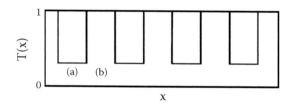

FIGURE 2.25 Transmittance (T) of halftone dot pattern vs. position (x). A dot is present at position (a) and the transmittance is low. There is no dot at position (b) so the transmittance is that of the paper.

location, x. Location (a) is an ink dot with T<1, and location (b) is the paper between dots where the transmittance is 1.

The point spread function and the halftone dot function are actually functions in terms of the two dimensions, P(x,y) and T(x,y). The theory for combining these two functions is based on a mathematical operation known as convolution, illustrated in Equation 2.23, where the symbol * is used as a shorthand notation for the convolution integral on the right side of Equation 2.23.

$$T(x,y) * P(x,y) = \int_{-\infty}^{\infty} \int_{-\infty}^{\infty} T(x,y) \cdot P(x-a, y-b) \, da \, db \qquad (2.23)$$

This convolution operation is combined with the reflectance of the paper, R_p, to calculate reflected pattern of light, R(x,y), from the halftone image. This is done with Equation 2.24. Note the * in the expression means the convolution integral, not multiplication.

$$R(x,y) = R_p \cdot T(x,y) \cdot \left[T(x,y) * P(x,y) \right] \qquad (2.24)$$

Semi-Empirical Model of Optical Dot Gain: Equation 2.24 is a concise way of expressing the physical optical dot gain, but the practical application of this equation can be very complex. Thus, simplifications have been suggested that provide useful approximations to the optical dot-gain mechanism. One example is a semi-empirical model based on a combination of optical theory and experimental measurements of micro-reflectance, illustrated in Figure 2.26.[23–25] The terms R_g and R_k represent the reflectance of the unprinted paper and of the ink printed at F = 1, respectively. The observation is that the reflectance values of the ink and paper do not remain constant at these values. Rather, they vary with changes in the dot area fraction. We represent these non-constant reflectance values as R_p vs. F and R_i vs. F in Figure 2.26.

The variation in the paper and ink reflectance values reflect the observed failure of the Murray–Davies Equation 2.7. When Equation 2.7 is used with constant values of R_g and R_k the result is a calculated predicted value of R is smaller than the experimental value. However, if the values R_p and R_i are used in Equation 2.7, where R_p and R_i are the values observed at each value of F, then the correct value of R is calculated. This means that the Murray–Davies equation is the correct equation to describe a halftone image, provided the correct values of values R_p and R_i are used instead of the constant values R_g and R_k.[23] The problem then is to model the way R_p and R_i vary with F. Such models have been developed and reported by several researchers, and the results have led to useful insights into the optical and physical behavior of halftone systems.

Mechanistic models such as those described above have contributed significantly to a better understanding of the printing processes,[23] and much work

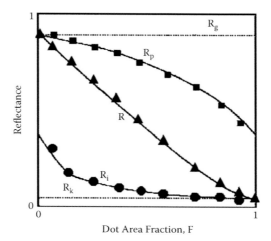

FIGURE 2.26 Example of micro-reflectance behavior of 30-LPI halftone images printed with a 600-dpi inkjet printer at dot fractions ranging between 0 and 1. R_i is the reflectance halftone dots, and R_p is the reflectance of the paper between the dots. R is the overall average reflectance of the image.

continues to be done in this area. However, routine calibration and control of printers and for practical color management statistical models are generally found to be more useful.

2.6 MAKING HALFTONE IMAGES

The method by which continuous tone images are converted into halftone images was first developed early during the 19th century concurrently with the development of photographic technology. During the late 20th century, digital techniques were developed to convert continuous tone images into halftone images. These digital techniques initially were designed to mimic the original photographic process. Thus, it is worthwhile to review the original photographic process for generating a halftone printing plate. This process is called pre-press process photography.[22]

2.6.1 PRE-PRESS PROCESS PHOTOGRAPHY

The basic photographic process for printing a film negative is illustrated in Figure 2.27. The key to the process is the relationship between the transmittance, T, or the film negative and the reflectance, R, of the photographic paper. This relationship is the tone transfer function of the print paper. Figure 2.28 illustrates a tone transfer function for a paper that produces a full range of copy gray values and one that produces a very high contrast copy with very few mid-tone grays.[22]

Process photography looks much like ordinary photography, but a special screen is added to convert grayscale to dots. This is illustrated in Figure 2.29. In

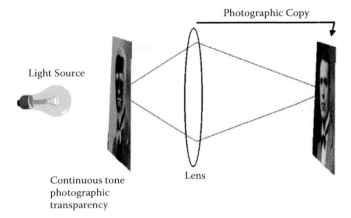

FIGURE 2.27 Printing a continuous tone film negative onto a continuous tone photographic paper. (Courtesy of Frank Cost.)

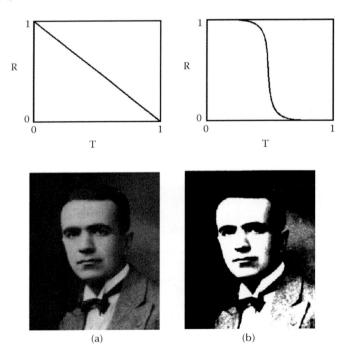

FIGURE 2.28 Examples of normal- and high-contrast tone transfer functions and the images they produce. (a) Normal contrast; (b) high contrast. (Courtesy of Frank Cost.)

addition, the photographic paper is replaced with a photographic film of very high contrast, such as in Figure 2.28(b).

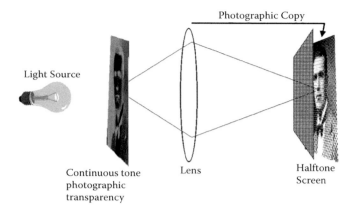

FIGURE 2.29 Halftone process photography. (Courtesy of Frank Cost.)

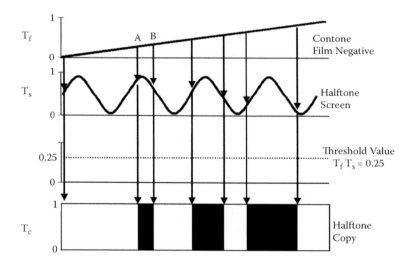

FIGURE 2.30 The effect of the halftone screen.

The high-contrast copy film acts as a threshold system that produces either black or white, but no gray values. A mid-tone gray in the original negative image (R = 0.5, for example) is at the threshold, so any value greater than that is copied as black (T = 0) and any value less than that is copied as white (T = 1). The halftone screen modulates the image, as illustrated in Figure 2.30, to produce dots of a size proportional to the gray value. The film and the light source are chosen, in this example, for a threshold gray value of 0.25. When the original film transmittance, T_f, times the halftone screen transmittance, T_s, is less than 0.25, there is too little light to have an effect on the copy film. Under that condition, the copy film remains at a high transmittance, $T_c = 0$ (white).

At point (A) in Figure 2.30, the negative film image has a transmittance of $T_f = 0.313$ and the screen has a transmittance of $T_s = 0.80$. The product is 0.25, so the copy film transitions to black, $T_c = 0$. Then at point (B) the original has $T_f = 0.35$ and the screen has $T_s = 0.714$ for a product of 0.25, so the copy film transitions back to white ($T_c = 1$). If we continue in this way, we note that the size of the region that is black increases as the transmittance of the con-tone film, T_f, increases. In other words, dots are produced with areas that are proportional to the con-tone gray value, T_f. Note that the size of the dot can be no greater than one cycle of the screen. The screen frequency, therefore, translates directly to the LPI halftone frequency of the image.

Both negative and positive types of process photography are practiced. Figure 2.30 is a negative process, just like regular photographic film. A positive halftone process involves a high-contrast copy film that turns clear on exposure and black in unexposed regions. In this case the black dots are where the white is in the halftone copy of Figure 2.30, and the white becomes black. Regardless of the technique, the result is a photographic transparency that is made of halftone dots rather than continuous tones of gray. This halftone film is then used as a mask to expose photosensitive material used to form the printing plate, as illustrated in Figure 2.31.

2.6.2 THE DIGITAL HALFTONING MASK

A digital simulation of the photographic process is often used to generate digital halftone images. This is done by dividing the original image into pixels at the same dpi as the printer, as illustrated in Figure 2.32. This process is called

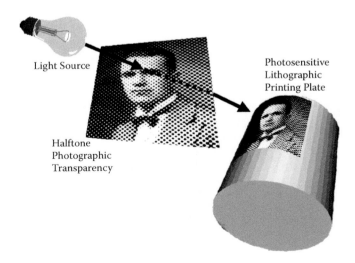

Light Source

Photosensitive
Lithographic
Printing Plate

Halftone
Photographic
Transparency

FIGURE 2.31 Exposure through the halftone photographic transparency to form an offset lithographic printing plate. (Courtesy of Frank Cost.)

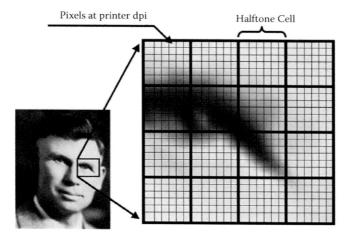

FIGURE 2.32 Digital halftoning begins by sampling the original image at the printer dpi and constructing digital halftone cells.

sampling, and the digital halftone is generated by an algorithm that determines whether to print ink or not to print ink at each pixel location. The halftone algorithm begins by grouping pixels into halftone cells. Each cell measures $N \times N$ pixels and will contain a digital halftone dot with a size proportional to the gray level in the original image. The cell size illustrated in Figure 2.32 is $N = 6$ for a 6×6 halftone cell.

Next, a matrix is designed as illustrated in Figure 2.33. This matrix, also called a digital halftone mask, is an $N \times N$ array of threshold numbers (t = 0, 1,

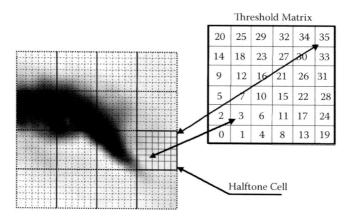

FIGURE 2.33 The halftone algorithm compares each halftone cell with the value, t, in the threshold matrix. The threshold matrix is also called a digital halftone mask.

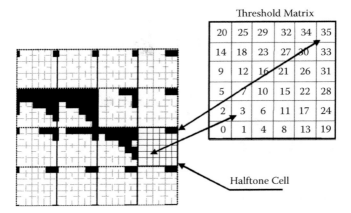

Threshold Matrix

20	25	29	32	34	35
14	18	23	27	30	33
9	12	16	21	26	31
5	7	10	15	22	28
2	3	6	11	17	24
0	1	4	8	13	19

Halftone Cell

FIGURE 2.34 If the gray value is less than the $t/(N^2)$ number divided by the matrix size, the pixel is black.

2 ... $N - 1$) used to determine whether or not ink is printed at each pixel location. This is done by comparing the gray levels in the original image with the corresponding threshold numbers.

For example, the halftone cell in the original image just to the right of the eye in Figure 2.33 is a uniform area of reflectance R = 0.92. This is compared to the value $t/(N^2)$ as illustrated in Figure 2.34. Thus, R = 0.92 is compared with 0/36, 1/36, 2/36 ... 34/36, 35/36. In all cases except the last two, R is larger, so the algorithm tells the computer to print no ink. In the last two cases. R < 34/36 and R < 35/36, so the algorithm tells the printer to print ink in the last two locations of the halftone cell. The resulting halftone does not look much like an eye, but when we zoom out to a reasonable distance, as illustrated in Figure 2.35, the appearance of gray scale becomes more apparent.

The quality of the halftone image is significantly influenced by the arrangement of the threshold numbers in the halftone cell. Figure 2.36(b) was generated using the threshold matrix shown in Figure 2.23. Figure 2.36(c) and (d) were done with the same 6 × 6 cell size but with the threshold matrices shown at the bottom of the figure. The halftone dots in (c) are more oval in shape and are much more pleasant in appearance. The threshold values for both (b) and (c) are clustered to form discrete dots. However, the threshold values for (d) are quasi-random. They form a dispersed halftone dot rather than a clustered halftone dot. In this illustration, the dispersed halftone dot makes the better-looking image.

In the early days of desktop printing, printers were limited in their addressability. A 100-dpi printer was considered to be a high-resolution printer for text imaging a decade ago. For such a printer to print halftones with 37 different gray levels, a 6 × 6 halftone cell was needed. This made a halftone cell of 6/100 inches or 6/4 = 1.5 millimeters in size. Such halftone cells were quite easy to see at an ordinary reading distance, so the design of the halftone cell was quite important to the quality of the printed image. For this reason, the dispersed type of halftone,

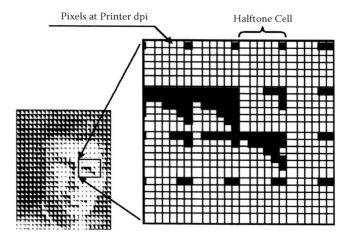

Pixels at Printer dpi   Halftone Cell

FIGURE 2.35 Zooming out from the halftone pattern makes the image look more like the original image.

as illustrated in Figure 2.36(d), was far superior to the traditional clustered dot halftone used by commercial printers. Other factors that influence image quality in digital halftone printing will be explored subsequently.

2.6.3 The Noise Distribution Technique: An Alternative to the Mask Technique

An alternative to the halftone mask technique for generating a digital halftone is a technique called noise distribution. Photographers and printers have long known that the addition of random or quasi-random noise to an image can improve the appearance of tone. Figure 2.37 is an example of stipple point engraving.[14] The artist used a tool to add random texture to the engraved image. The result is a fine set of random ink dots that are more closely spaced in shadow regions and more widely spaced in highlight regions.

Robert's Method: One of the earliest digital techniques for using noise to simulate grayscale was developed by L. G. Roberts in 1962.[26] Robert's method is illustrated in Figure 2.38(a).

This technique involves sampling an original image as described in Figure 2.32, but the sampled pixels are not combined into halftone cells. Instead, the decision about whether or not to print ink at a pixel location is based on a comparison between the original gray value and a random number, Rnd, between 0 and 1. For example, if $R = 0.5$, there is a 50% change that $R > Rnd$. Note also that $(1 - R) < Rnd$ 50% of the time. If we print ink when $(1 - R) < Rnd$, then we will print ink 50% of the time. Similarly, if $R = 0.2$, then $R < Rnd$ only 20% of the time, and $(1 - R) < Rnd$ 80% of the time, so 80% of the pixels are printed with ink.

Floyd–Steinberg and Error Diffusion: In 1975 Floyd and Steinberg introduced a halftoning technique they called an adaptive algorithm.[27] This technique

FIGURE 2.36 Three halftone patterns.

distributes quasi-random noise in the image in a way that significantly improves the visual quality of the image.

The technique starts with the first pixel at location $(i,j) = (0,0)$ in the upper left corner of the image and compares the original gray value at that location, $RO_{0,0}$, with a threshold value of 0.5. If $(1 - RO_{0,0}) < 0.5$, then ink is printed;

FIGURE 2.37 Since the Renaissance, printers have applied a variety of techniques to induce random textures to mid-tone gray regions of printed images.[14] This is an example of a technique called stipple point engraving. (Courtesy of Melbert E. Cary, Jr. Graphic Arts Collection, Rochester Institute of Technology.)

(a)   (b)

FIGURE 2.38 Robert's method of error addition and Floyd–Steinberg's method of error diffusion. (a): Robert's method; (b): Floyd–Steinberg method.

otherwise, ink is not printed. This first pixel in the copy image should therefore have an ideal gray value of $RC_{0,0} = 0$ or 1. The difference between the original and the ideal copy reflectance ($E_{00} = RO_{00} - RC_{00}$) is a measure of the amount of error that occurs in the threshold process. For example, if $RO_{00} = 0.2$, then $RC_{00} = 0$, so $E_{00} = 0.2 - 0 = 0.2$, and the pixel is too dark by 0.2. To correct for this error, we propagate this error to the next pixel. We do this by adding the error to the threshold value of the next pixel. Thus, the value of $E_{00} = 0.2$ is added to the threshold value for the next pixel at location $(i,j) = (0,1)$. Using the new, adjusted threshold, we examine whether or not $(1 - RO_{0,1}) < 0.5 + 0.2$. If it is,

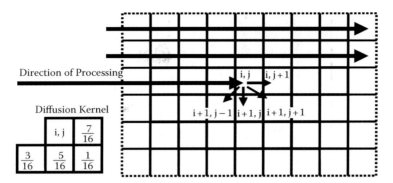

FIGURE 2.39 The traditional Floyd–Steinberg diffusion kernel.

then we print ink; otherwise, we do not. Note that by propagating the error, we increase the probability that ink will not be printed because the threshold value is increased.

The error can be either positive or negative, so the threshold value can be increased or decreased. In addition, we generally do not propagate the error only to the next pixel. We diffuse a fraction of the error to several neighboring pixels. The most often used pattern for error diffusion is illustrated in Figure 2.39.

In this example, the total error, $E_{i,j}$, made in the thresholding process at any location (i,j), is divided up and added to four other threshold values at four other locations. Then $(7/16)E_{i,j}$ is added to the threshold value that will be used for pixel (i,j+1); $(3/16)E_{i,j}$ is added at location (i+1,j-1); $(5/16)E_{i,j}$ is added at (i+1,j); and $(1/16)E_{i,j}$ is added at (i+1,j+1). Figure 2.38(b) is an example of the Floyd–Steinberg technique of error diffusion.

2.6.4 CLASSIFICATION OF DIGITAL HALFTONING

The two major techniques for generating digital halftones may be classified as shown in Table 2.6. These are digital masking, as described in Section 2.6.2, and noise distribution, as described in Section 2.6.3. This is an arbitrary classification, and as with any classification scheme, there are cases reported in the research

TABLE 2.6
Techniques for Digital Halftoning

Techniques for Digital Halftoning	Example #1	Example #2
(A) Masking	Clustered Dot	Dispersed Dot
(B) Noise Distribution	Robert's Method	Error Diffusion

(a)   (b)

FIGURE 2.40 Illustration of spatial frequency. (a): A low-frequency image; (b): a high-frequency image.

literature that do not clearly fit one or the other category. Nevertheless, halftone masking and noise distribution are the major techniques in common use.

Digital halftoning has developed rapidly over the past 3 decades, and this has led to confusing names for the different techniques. For example, the term *dither* is often used to describe noise distribution techniques such as error diffusion and Robert's method. Unfortunately, the term *dither mask* is often used to refer to the digital halftone mask (threshold matrix) described in Section 2.6.2. As a result, *dithering* has evolved to mean any type of digital halftoning and thus has lost its utility for distinguishing between different techniques for generating halftones.

Another classification scheme is based on characteristics of the halftone pattern rather than the technique used to generate it. The AM/FM classification scheme is an example. All halftones control grayscale by controlling the dot area fraction, F, but there are two different kinds of patterns for controlling F. One pattern, called the AM pattern, is illustrated in Figure 2.36(b) and (c). The AM halftone varies F by varying the size of the halftone dot. The other pattern is called the FM pattern and is illustrated in Figure 2.36(c) and Figure 2.38(a) and (b). In these halftones, F is varied by varying the number of printer dots in a unit area, but the dots do not change in size. Note that the halftone mask technique can produce either AM- or FM-type halftone patterns. The noise distribution technique produces only FM halftones.

More recent work has shown that the concept of AM vs. FM halftones represents two extremes and that many halftones have attributes of both kinds. The respi halftone, shown in Figure 2.40, was designed to change from 220 LPI to 110 LPI in the highlights.[28] Most printing devices print large dots more reliably than small dots. When the size of the AM dots at 220 LPI becomes too small for the printing device to print reliably, a frequency change (FM effect) down to 110 LPI allows highlights to be printed with fewer large dots.

TABLE 2.7
Conversion Table between Frequency and Size.
Frequency Is in LPI (line pairs per inch) and cy/mm (cycles per millimeter). Size Is in Mils (1/1000 inch) and Millimeters

LPI	mils	cy/mm	mm
12	0.83	0.472	0.021
35	28.6	1.378	0.726
72	13.9	2.835	0.353
85	11.8	3.346	0.299
100	10	3.937	0.254
110	9.1	4.331	0.231
120	8.3	4.724	0.212
150	6.7	5.906	0.169
220	4.5	8.661	0.115
300	3.3	11.811	0.085
600	1.7	23.622	0.042
1200	0.8	47.244	0.021
2400	0.4	94.488	0.011

A more general, but more complex, technique for characterizing halftone patterns involves measuring the so-called noise power spectrum of the halftone image. This provides a more general understanding of the halftone pattern than the simple concept of AM vs. FM. Understanding the noise power spectrum is very helpful for understanding the image quality differences between different halftone patterns.

2.7 IMAGE QUALITY AND THE NOISE POWER SPECTRUM

The concept of spatial frequency is illustrated by the two images in Figure 2.40. Frequency refers to how often or how quickly something happens in an image.

In Figure 2.40(a), only one person is shown, but in image (b) people are shown. People appear in (b) more frequently than they occur in (a), so in terms of people, (b) is a higher frequency image than (a). This also means the people in (b) are smaller than in (a), with both images completely filled with people. As the frequency of an image element increases, its size decreases. Halftone dots are another example. A 35-lpi halftone is composed of halftone cells that are 0.0286 inches (28.6 mils) diameter, and a 100-lpi halftone has cells of 10 mils.

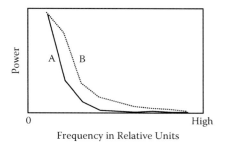

FIGURE 2.41 Noise power spectra of images in Figure 2.40(a) and (b). The axis units are relative units for illustration purposes.

The lpi is an index of halftone frequency. The relationship between frequency and size is summarized in Table 2.7.

Frequency is a useful concept for describing how rapidly individual elements of an image change. Consider the element "people" in Figure 2.40a. The number of people changes slowly as we move across the image, but in (b) the people change rapidly. Another example is the element "edge." For example, the edge of the man's collar is a high-frequency element because it changes rapidly as we move from left to right over the edge. Sharper edges are high frequency, but blurred broad edges are low frequency. Frequency is a useful metric of image quality for describing the intrinsic resolution characteristics of an image.[29]

2.7.1 FOURIER ANALYSIS OF NOISE POWER

An 18th-century mathematician named Jean Baptist Joseph Fourier demonstrated that any shape can be described as the sum of all of the frequencies in the image. He also developed the mathematical process for extracting all the frequencies in the shape. This process is called the Fourier transform of the image and is done easily with mathematical applications such as spreadsheets.[29] As an example, the Fourier transforms of the images in Figure 2.40(a) and (b) were calculated and are plotted in Figure 2.41. The horizontal axis is the frequency and the vertical axis is the square of the magnitude of the Fourier transform. It represents the relative importance of image features at each frequency. Thus, the Fourier analysis shows, in a quantitative way, that image (b) has more power at higher frequency. In other words, (b) is a busier image than (a).

2.7.2 NOISE POWER AND THE HUMAN VISUAL SYSTEM

Halftoning is a very useful way to control not only the gray level in an image, but also the noise in an image. Halftones are a type of granularity pattern and can be an objectionable source of noise in the image. To minimize the visual impact of this halftone noise, we prefer to print dots that are smaller than the eye can see. The resolving power of the human visual system reaches a maximum for features of about 0.8 to 1.0 mm in size for a normal reading distance of 35 cm. This means we are most sensitive to features about the size of printed letters. Features that are approximately 0.23 mm are on the threshold of resolution of normal human vision. As shown in Table 2.2, this corresponds to a 110-LPI halftone. A 150-LPI clustered dot halftone uses halftone cells that are not resolvable at all by most people at a normal reading distance. For this reason, most magazine images are printed at 150 LPI in order to look indistinguishable from a continuous tone photograph.

Desktop printing in the 1980s involved devices capable of only limited addressability. A simple dot matrix printer, for example, with an addressability of 72 LPI could construct letters and numbers of reasonable quality. However, a 6 × 6 halftone cell able to print 37 levels of gray resulted in halftones cells 2.1 mm on a side. Figure 2.42(c) is an illustration of a 25% gray level printed in this way, and the halftone pattern is quite visible and objectionable. The other patterns in Figure 2.42 also illustrate a 72-LPI printer but with different halftone patterns. Some are clearly preferable to others.

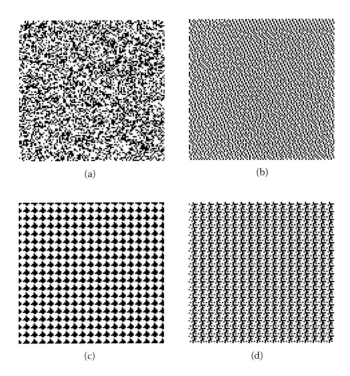

(a)   (b)

(c)   (d)

FIGURE 2.42 A gray level of R = 0.40 printed with a 72-dpi printer using four different halftone patterns. (a): Roberts method, (b) Floyd–Steinberg method. (c) and (d) are 6 × 6 mask halftones made with the matrices shown in Figure 2.36.

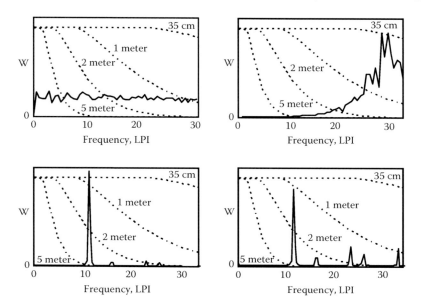

FIGURE 2.43 Noise power spectra of the halftone patterns shown in Figure 2.42. Dotted lines are the resolution of the human visual system at viewing distances of 0.35, 1, 2, and 5 meters.

Fourier analysis provides an important technique for measuring the severity of noise patterns of different types. Figure 2.43 shows the Fourier noise power spectra for the patterns in Figure 2.42. One can think of the noise power, W, as the statistical variance (squared standard deviation) of the reflectance in the image measured for each frequency contained in the image (frequency = 1/size of image element). Both of the masking techniques of Figure 2.42(c) and (d) show large peaks at 0.47 cy/mm (12 LPI), which is the frequency of the halftone cells. However, the dispersed dot image of Figure 2.42(d) looks better to most people. Part of the reason it looks better is that some of the noise power has been shifted from the dominant 12 LPI into new peaks at higher frequency, where our eyes are less sensitive. Robert's method in Figure 2.42(a) uses white noise to simulate gray scale. As the name implies, white noise contains equal noise power at all frequencies. The best-looking pattern is in Figure 2.42(b), the Floyd–Steinberg pattern. The Floyd–Steinberg process shifts most of the noise to high frequency (small feature sizes) so the eye does not resolve it as well.

The usefulness of noise power as a method of evaluating granularity in printed images can be illustrated by examining Figure 2.42 at different reading distances. If we slowly back away from Figure 2.42, we will notice a distance at which the Floyd–Steinberg image of Figure 2.42(b) blends to a uniform gray. For most people, this occurs at about 2 to 3 meters of viewing distance. As shown in Figure 2.43, the sensitivity of the human visual system at a distance of 2 meters drops to nearly zero at 1 cy/mm. Because most of the noise in the Floyd–Steinberg image is above 1 cy/mm, the noise in Figure 2.43 nearly disappears at 2 meters viewing distance.

By slowly moving back farther than 3 meters, one will notice that the noise power of the other images decreases. The clustered dot and dispersed dot images show a sharp decrease in visual noise at about 4 to 6 meters of viewing distance. Most people observe that the dispersed dot image of Figure 2.42(d) blends to a noiseless gray at a slightly closer distance than the clustered dot image of Figure 2.42(c). Again, this is a result of shifting some of the noise power to higher frequencies. Because Robert's method uses white noise with noise power present equally at all frequencies, it retains noise power at all viewing distances.

2.7.3 OBJECTIVE MEASURES OF GRANULARITY CONSTANTS

It is convenient to define a single number to use as a comparative index for the granularity of different images. Of course, a single number index is insufficient to capture all of the facets of image noise and the visual impact of regular vs. random patterns. However, a single index of granularity can be useful if we remember that it is only part of the story.

A simple index of visual granularity is the minimum viewing distance at which an image blends to a constant gray, as described above. This technique can be carried out quite well by a single observer, and the results are quite useful for the comparison of different halftone patterns. This technique was one of the first quantitative measurements of granularity used by the photographic industry to measure the granularity of silver halide photographs.[30]

The most common instrumental technique for measuring the granularity of printed halftone images is to capture a digital copy of the printed image with a camera or scanner that has a resolution well beyond the dpi addressability of the printer. Once the image of the printed sample is captured, a granularity constant can be calculated by digital image analysis. The most common index of granularity is the standard deviation of the captured image, symbolized σ. However, more is required than just a captured image. The image capture device should be calibrated, and it is important to know whether the imaging device is calibrated to reflectance, R, or to reflection density, $D = -\log(R)$ or to some other function of reflectance. The standard deviations σ_R and σ_D are quite different.

Equation 2.25 and Equation 2.26 are used to calculate σ_R and σ_D; $\bar{R}$ and $\bar{D}$ are the average values of reflectance and density of the image; R_j and D_j are the individual reflectance and density at each pixel location, j, in the image; and N is the total number of pixels in the captured image. The values of σ_R and σ_D are sometimes called root mean squared (RMS) granularity constants.

$$\sigma_R = \sqrt{\frac{1}{N} \cdot \sum_{j=1}^{N} \left(R_j - \bar{R}\right)^2} \qquad (2.25)$$

$$\sigma_D = \sqrt{\frac{1}{N} \cdot \sum_{j=1}^{N} \left(D_j - \bar{D}\right)^2} \tag{2.26}$$

Ideal printed samples of all of the halftones in Figure 2.42 have the same RMS granularity, given by Equation 2.27. This is because the calculation of granularity does not take into account the filtering that occurs by the human visual system.

$$\sigma_R = F \cdot \left(1 - F\right)\left(R_p - R_i\right) \tag{2.27}$$

To account for the filtering effect, we need an equation for the human visual system. Equation 2.28 is an approximate equation for describing the spatial sensitivity function of the human visual system, the dotted line in Figure 2.43(A), for a 35-cm reading distance. We call this function the visual transfer function (VTF). Notice that frequency in both the horizontal and vertical directions, ω, v, is involved in the complete analysis.

$$\text{for } \sqrt{\omega^2 + v^2} > 1 \text{ cy / mm}, \text{VTF}\left(\omega, v\right) = 5.05 \cdot \left(e^{-0.84\sqrt{\omega^2 + v^2}} - e^{-1.45\sqrt{\omega^2 + v^2}}\right)$$

$$\text{for } \sqrt{\omega^2 + v^2} \le 1 \text{ cy / mm}, \text{VTF}\left(\omega, v\right) = 5$$

$$\tag{2.28}$$

where cy means cycles.

The image has two directional dimension, $R(x,y)$, so the complete Fourier analysis actually leads to a two-dimensional noise power spectrum, $W(\omega, v)$. The filter calculation is performed with the integral Equation 2.29. The result, σ_V^2, is the squared RMS granularity that the eye is able to see.

$$\sigma_V^2 = \iint_{\dot{u}, i} W\left(\dot{u}, i\right) \text{VTF}\left(\dot{u}, i\right) d\dot{u}\, d\, i \tag{2.29}$$

Integral Equation 2.29 does two things. First, it is the multiplication of the noise power spectrum, W, and the VTF. The result of this multiplication is shown in Figure 2.44 for Robert and Floyd–Steinberg images from Figure 2.42. The integration process simply calculates the area under the curve. Thus, the square root of the area under each curve is a measure of the RMS image deviation, σ_V, which the eye can detect at the viewing distance shown. Recall that σ from Equation 2.25 is the same for all four images of Figure 2.42 at all reading distances. However, σ_V changes with both reading distance and with the halftone

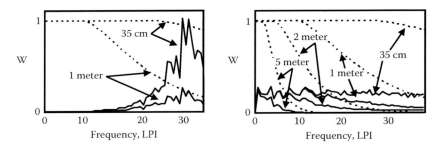

FIGURE 2.44 Illustration of the attenuation of noise power by the human visual system at various viewing distances for Roberts and Floyd–Steinberg methods, Figure 2.42(a) and (b).

pattern. Thus, σ_V correlates with our visual experience of observing different halftone patterns at different viewing distances.

2.7.4 GRANULARITY AND THE PRINTING DEVICE

The filtering process illustrated with Equation 2.29 is an example of noise filtering in general. Many things can filter out noise, and the process of printing an image is one example. In the above discussion, a perfect printing process was assumed. That is, it was assumed that the halftone dots are perfectly sharp edges. However, the dot-gain effects described previously involve spreading and blurring, and these effects can reduce the noise power, as illustrated in Figure 2.45.

Another printer effect that often plays an even more important role in the quality of halftone printing is the stability of the printing process itself. In addition to blurring the noise that is intrinsic to the halftone pattern, the printing process can contribute noise of its own. Figure 2.46 illustrates a printing process that introduces both random noise and a regular banding pattern. Both types of noise are examples of what is often called printer instability.

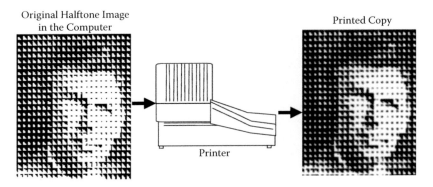

FIGURE 2.45 The printing process can blur the halftone dots and reduce noise.

Original Halftone Image
in the Computer   Printed Copy

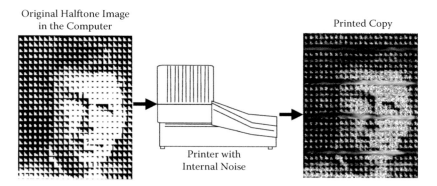

Printer with
Internal Noise

FIGURE 2.46 The printing process can blur input noise and add noise of its own. Random and banding noise is illustrated.

The degree of instability manifested by the printing process often depends on the pattern that is printed. Some patterns are easier to print than others are. In particular, many printing devices show more printer instability when attempting to print small dots than when printing large dots. For this reason, it is often better to print a traditional clustered dot halftone than an FM pattern such as Floyd–Steinberg. This is particularly true for printers capable of addressabilities above 600 dpi. A 6 × 6 clustered dot halftone would produce a pattern of dots that is barely visible at an ordinary reading distance. However, a Floyd–Steinberg image in the presence of significant printer instability may actually be more objectionable than a uniform, stable clustered dot image. For this reason, FM halftones are rarely preferred over clustered dots for high dpi printers.

REFERENCES

1. P.G. Engeldrum, *Psychometric Scaling: A Toolkit for Imaging Systems Development*, chapt. 2, Imcotek Press, Winchester, MA, 2000.
2. R.R. Buckley, The History of Device Independent Color — Ten Years Later, Tenth Color Imaging Conference: Color Science and Engineering Systems, Technologies, Applications, Scottsdale, AZ, November 12, 2002, pp. 41–46.
3. P.A. Crean and R. Buckley, Device independent color: Who wants it? in K. Braun and R. Eschback, Eds., *Recent Progress in Color Science*, IS&T, Springfield, VA, 1997, pp. 230–232.
4. R. Brook and G. Arnold, *Applied Regression Analysis and Experimental Design*, Marcel Dekker, New York, 1985.
5. W. Thomas, Jr., Ed., *SPSE Handbook of Photographic Science and Engineering*, chapt. 17, John Wiley & Sons, NY, 1973.
6. USAF Test Target, U.S. Government specification Mil-Std-150.
7. NPS 1963A Microcopy Target. Also known as NBS 1010A Microcopy Test Chart and ANSI/ISO Test Chart #2.

8. P.G.J. Barten, *Contrast Sensitivity of the Human Eye and Its Effects on Image Quality*, SPIE Press Monograph Vol. PM72, Bellingham, WA.

9. F.D. Kagy, *Graphic Arts Photography*, Delmar Publishing, NY, 1983.

10. R. Berns, *Principles of Color Technology*, 3rd ed., chapt. 2, p. 31, John Wiley & Sons, NY, 2000.

11. R.W.G. Hunt, *Measuring Colour*, 2nd ed., Ellis Horwood, London, 1995.

12. R.W.G. Hunt, *The Reproduction of Colour*, Fountain Press, England, 1987.

13. J.A.C. Yule, *Principles of Colour Reproduction*, John Wiley & Sons, NY, 1967.

14. B. Gascoigne, *How to Identify Prints*, p. 54, Thames and Hudson, NY, 1991.

15. F. Cost, *Pocket Guide to Digital Printing*, Delmar Publishers, NY, 1997.

16. H.E.J. Neugebauer, Die Theoretischen Grundlagen des Mehrfarbenbuchdrucks, *Z. Wissenschaft. Photograph. Photophys. Photochem.*, 36, 22, 1937.

17. P.G. Engeldrum, The color between the dots, *J. Imag. Sci. Technol.*, 38, 545, 1994.

18. E.J. Giorgiann and T.E. Madden, *Digital Color Management*, Addison-Wesley, Reading, MA, 1998.

19. J.A.S. Viggiano, Modeling the Color of Multi-Colored Halftones, *TAGA Proceedings*, TAGA, Vancouver, BC, Canada, 1990.

20. T.N. Pappas, *Digital Halftoning Techniques for Printing*, IS&T 47th Annual Conference, p. 468, 1994, and references contained therein, IS&T, Springfield, VA.

21. T.N. Pappas, C.-K. Dong, and D.L. Neuhoff, Measurement of printer parameters of model-based halftoning, *J. Electronic Imag.*, 2, 193, 1993.

22. G.L. Rogers, Optical dot gain: lateral scattering probabilities, *J. Imag. Sci. Technol.*, 42, 341, 1998.

23. J.S. Arney, P.G. Engeldrum, and H. Zeng, An expanded Murray–Davies model of tone reproduction in halftone imaging, *J. Imag. Sci. Technol.*, 39, 502, 1995.

24. J.S. Arney and M. Katsube, A probability description of the Yule–Nielsen effect, *J. Imag. Sci. Technol.*, 411, 633 and 637, 1997.

25. J.S. Arney, T. Wu, and C. Blehm, Modeling the Yule–Nielsen effect on color halftones, *J. Imag. Sci. Technol.*, 42, 335, 1998.

26. L.G. Roberts, Picture coding using pseudorandom noise, *IRE Trans.*, IT-8(2), 145–154, 1962.

27. R.W. Floyd and L. Steinberg, Adaptive algorithm for spatial gray scale, *SID Int. Sym. Dig. Technol. Papers*, pp. 36–37, 1975.

28. M.H. Bruno, *Imaging for Graphic Arts, Image Processes and Materials*, p. 472, Van Nostrand Reinhold, NY, 1989.

29. M.A. Kriss, Image structure, in T.H. James, Ed., *The Theory of the Photographic Process*, 4th ed., MacMillan, NY, 1977.

30. C.E.K. Meese, *The Theory of the Photographic Process*, 1st ed., chapt. 21, MacMillan, NY, 1941.

3 The Business and Market for Desktop Printers

Frank Romano

CONTENTS

3.1 HISTORICAL

In 1878, James Clephane, a stenographer and developer involved in the typewriter and the linotype typesetter, said, "I want to bridge the gap between the typewriter and the printed page." [1] There is no doubt that the current breed of digital printers provides a performance level equal to low-level printing presses. The concept of desktop publishing is based on the ability to compose typographic pages and output them at a quality level competitive with the graphic arts. The typewriter went down one road to business communication and the linotype went down another road to the printing industry. It would be the digital printer where both roads would meet.

The desktop printer market began because Robert Howard wanted to personalize poker chips. He sought a method to print around the periphery of a gambling chip and came up with the concept of impacting the surface of the chip through a ribbon with small pins that would image the letters, very much as a television image was composed of raster lines. (Howard had been a pioneer in TV manufacturing.) He and Dr. An Wang became partners in a company to market an early digital computer co-invented by both. When they decided to separate, Wang formed Wang Laboratories and Howard founded Centronics Data Computer Corporation in 1969. Howard developed the first dot matrix printer, and Centronics became one of the world's leading computer printer manufacturers. Today's popular laser printers originated from a subsequent Canon/Centronics joint development program in which Mr. Howard served as a seminal innovator. Howard served as chairman of Centronics from 1969 to 1982, when Control Data Corporation acquired a controlling interest.

The Centronics parallel plug became the single common denominator in the computer world. Howard tells the story of how it came to be. "Dr. Wang had ordered a large number of these connectors for another project. Since there were so many we decided to use it for the printer cable. The rest is history." The desktop dot matrix printer was an important enabler in the evolution of desktop computing. A computer printer produces hard copy (text or graphics usually on paper) from data stored in a computer connected to it. The first of the printers were connected via telephone lines to mainframe computers. They were used by hotels, airlines, rental agencies, and other organizations.

As mainframe computers, and then minicomputers, came into use, there was a need to print their output. The first computer printers were typewriter adaptations. They evolved into higher- and higher-speed printers, all using impact and ribbon technology. These printers may be broadly characterized as:

- Character printer
  - Typewriter printer
  - Daisywheel printer
  - Chain or band printer

- Dot matrix printer
  - Character matrix printer
  - Line matrix printer
  - Page printer
  - Thermal printer (used in fax devices)

There are two major mechanisms used for dot matrix printing:

- Ballistic wire printer
- Stored energy printer

Ballistic wire computer printers have a printhead with holes drilled through it. Thin wires go through the printhead and pawls are actuated by solenoids. The pawls strike the wires, causing them to go out and hit an inked ribbon, which hits the paper. After about a million characters, even with tungsten blocks and titanium pawls, the printing becomes too unclear to read. The print resolution ranges from a low of about 50 dots per inch (for receipt printers) to a high of about 300 dots per inch (better than the human eye can see at 14 inches), used on premium graphic printers.

A stored-energy printer is a type of computer printer that uses the energy stored in a spring or magnetic field to push a hammer through a ribbon to print a dot. These printers print millions to billions of dots per hammer. The most common printers to use this have been the line-matrix printers made by Printronix and its licensees. In these, the hammers are arranged as a hammerbank, a sort of comb that oscillates horizontally to produce a line of dots. In a character matrix printer, the hammers are machined from an oval of magnetically permeable stainless steel, and the hammer tips form a couple of vertical rows. The original technology was patented by Printronix in 1974.

Although character printers provided higher quality, most development was applied to dot matrix printers. 5×7, 9×14, 11×16, and other matrices evolved to increase the quality and allow a larger character set. In 1979 the Sanders all-points addressable printer tried to bridge the gap between the limitations of mechanical pin technology and the emerging world of non-impact printing.

Throughout the 1970s, the quest for higher speed, larger character sets, and the ability to handle graphics led to the development of non-impact printing systems. These can use any of the following technologies:

- Laser (toner) printer
- Inkjet printer
- Dye-sublimation printer
- Thermal printer

The introduction of the desktop computer in the late 1970s accelerated the need for desktop printers. Most were still impact printers at this time. Printers were available for use with home computers at prices of more than $1000. Most printers offered 96 characters in the standard American Standard Code for Information Interchange (ASCII) set. There were two main types of printers available: character printers and dot matrix printers.

Character printers operated like a typewriter by striking a piece of metal type against a ribbon and onto the paper. This type of printer was often called an impact or letter quality printer. It used either a type ball, such as IBM's Selectric typewriters, or a wheel with spokes that radiated out from the center, with the type characters at the end of the spokes, also called a daisy wheel.

Dot matrix printers formed characters with a series of pins in a vertical row that struck the ribbon and produced dots on the paper. As the printhead moved across the paper, the dots were printed in patterns that formed letters and numbers.

The matrix used to form a character was usually referred to as the number of horizontal dots by the number of vertical dots. A 5×7 matrix, for example, used up to five dots across and up to seven dots down.

The Centronics 730 was the first standard printer. It used a parallel cable whose pin layout went on to also become a standard for use with personal computers. That pin layout on parallel cable plugs is still in use today. Centronics also had several other models, including the 737 and 739. A less expensive printer made by Centronics, the 779, used 5×7 dot matrix characters and could print in sizes from 10 to 16.5 characters per inch (cpi), ranging from 60 characters per second (cps) at 10 cpi to 100 cps at 16.5 cpi. It also had a one-line buffer (which held up to 132 characters) but printed a limited 64-character ASCII set, all uppercase plus some special characters.

A company named Trendcom used mechanical solenoids that drove heated pins into a printhead — these were thermal printers that needed a special heat-sensitive paper. Their operation was very quiet, about as loud as sliding a finger across a piece of paper. The Model 100 printed 40 characters per line on paper that was about 4 1/2 inches wide. The Model 200 could print 80 columns per line on paper 8 1/2 inches wide. The first printer offered by Radio Shack for its TRS-80 computer also used a thermal printer.

Epson began in the printer business with the Epson MX-80, one of the first dot matrix printers that sold for less than $1000. A later version of this printer, the Epson MX-100, became available in early 1982. The MX-100 was a wide carriage model and could print high-resolution (hi-res) graphics without the need to add any special hardware. Epson printers were unique because they had a special feature called a double print mode in which a line was printed normally and then the paper was advanced 1/216 of an inch and the same line was printed again. This filled some gaps between dots on individual letters and made printouts more pleasing to the eye. In print enhancement mode, the pins hit the ribbon harder and made it possible to make multiple copies using carbon paper.

After the Silentype printer was released in 1979, Apple looked for another printer that would produce better, more permanent output than could be achieved with a thermal printer. One of the main problems with thermal paper was that, with time, the printing could fade. The Apple Dot Matrix Printer was released in 1982 for $699. Made from a modified C. Itoh printer, it was one of the first dot matrix printers that sold for under $1000. The Apple ImageWriter, released in 1983, was also made by C. Itoh; it had a faster print speed (120 cps) and could print in eight different pitches (character widths). It was very reliable and sold originally for $675. Later, a wide carriage version whose abilities were otherwise identical was made available. It was replaced by the ImageWriter II in September 1985.

In 1984, Hewlett Packard introduced the LaserJet laser printer based on the Canon LBP-CX 8-ppm (pages per minute) 300-dpi (dots per inch) engine. This was a significant breakthrough in printer quality and was capable of producing documents that looked professionally typeset. Apple decided to develop its own laser printer and in 1985 released the LaserWriter. The LaserWriter was supported only on the Macintosh and did its work through a page description language called PostScript. Apple entered the inkjet printer market in 1991 when it released the abortive Apple StyleWriter.

3.1.1 EPSON

Epson was established in 1961 under the name of Shinshu Seiki (Shinshu Precision Manufacturing Company) to provide precision parts for Seiko watches. The company was awarded a contract to make precision timers for the 1964 Olympics and also picked up work to build a printer. Thus, the EP-101 printer came about, and in 1968 it became one of the first printers for electronic calculators to hit the commercial market. Until the 1980s, approximately 100,000 of these printer mechanisms were being produced each month — about 90% of the world market.

Electronic watches using relatively high-current light-emitting diodes first appeared on the market in 1970, and Shinshu Seiki began researching a low-current alternative. It came up with the liquid-crystal display in 1974, which, in turn, led to LCD watches. Epson America was incorporated in 1975, with offices set up in Torrance, California, to distribute Shinshu Seiki products.

Epson's first dot matrix printer, TX-80, was introduced in 1978. However, this didn't attract much attention except from Commodore, which used it as the system printer for its PET computer. (The "80" in "TX-80" refers to the number of columns it printed per line.) An improved version, the MX-80, began development later that same year. The TX-80 took 3 months to develop. The MX-80 took about 2 years, was introduced in late 1980, quickly became the best-selling printer in the United States, and eventually became the industrial standard for microcomputers. This occurred despite its being designed not to produce graphics. Within a year, the Graftrax version, with graphics, had hit the streets.

3.1.2 CANON

Canon had the advantages of large investments and experience in copier technology. The company was able to leverage this base into fax machines, copiers, and later printers. In the latter half of the 1970s, engineers at Canon's Product Technology Research Institute conducted research on printing technologies for the next generation of copying machines. Initial work was devoted to producing piezo-elemental data necessary for inkjets, but this pursuit led to the discovery of a new technology. During testing, a hot soldering iron accidentally touched the needle of an ink-filled syringe, causing ink to spray from the needle's tip. Witnessing this, a member of the research team realized that heat, instead of pressure, could perhaps be used to induce the spray of ink. In 1997, technical concepts born of this discovery were combined with thermal-head technologies under development at the time. This enabled the number of ink nozzles to be multiplied, a notion that was previously inconceivable, opening the way to a new high-speed printing technology.

Numerous tests and refinements were made until Canon succeeded in developing the world's first Bubble Jet printing method in 1981. Canon continued to make refinements, finally unveiling the BJ-80 Bubble Jet printer 4 years later. This huge technological development was the product of 8 years — the period from the discovery of the initial principle to commercialization.

3.1.3 HEWLETT PACKARD (HP)

HP became a major player in the computer industry in the 1980s with a full range of computers, from desktop machines to portables to powerful minicomputers. HP made its entry into the printer market with the launch of inkjet printers and laser printers that connect to personal computers. The HP LaserJet printer line debuted in 1984 to become the company's most successful single product line ever.

In the 1980s, HP fueled the desktop publishing revolution with the introduction of high quality and reliable laser and later inkjet printers that connect to personal computers. The LaserJet was its most successful single product line ever and the DeskJet inkjet printers spelled the death of dot-matrix printers.

In the 1990s, HP became a follower in the desktop printer market because the company was slow with a color LaserJet, it used its own page description language instead of PostScript, and its printers were expensive. The desktop printer market is extremely competitive, and Epson and Canon gained ground in the inkjet market at the expense of HP.

HP announced a $1.2 billion investment in the imaging and printing market, including a planned roll out of 50 new products, in 2003. In 2005, HP announced a $2 billion investment in new printer technology and over 100 new products. HP is capitalizing on the ever-expanding digital photography imaging market. According to analysts, billions of digital images will need to be printed and inkjet printing will be the technology of choice.

HP has finalized its acquisition of Indigo N.V., an Israel-based manufacturer of specialty and commercial digital color presses. The digital printing market has been maturing for a decade, and given its successes with LaserJet technologies, HP would be a logical contender in the arena. By acquiring Indigo, which has established digital presses and an established customer base, the company has rolled out a centralized workflow management tool called HP Production Flow as well as a Web-based integrated publishing tool called HP Custom Publishing.

3.2   COMPUTER PRINTING EVOLVES

In the early days of the personal computer (PC), printing was simple. The PC owner bought a cheap printer, usually a dot matrix that minimally supported ASCII, and plugged it into the computer with a parallel cable. Applications would come with hundreds of printer drivers to output DOS (Disk Operating System) or ASCII text. The few software applications that supported graphics generally could only output on specific makes and models of printers. Shared network printing, if it existed, was usually done by some type of serial port (A-B boxes) switchbox.

When the Windows operating system was released, application programmers were finally free of the restrictions of how a printer manufacturer would change printer control codes. Graphics printing, in the form of fonts and images, was added to most applications, and demand for it rapidly increased across the corporation. Large, high-capacity laser printers designed for office printing appeared on the scene. Printing went from 150 dpi to 300 dpi to 600 dpi for the common desktop laser printer.

Inkjet printers have become widespread only in the past decade, yet the technology has been under development for more than 50 years. Inkjet recorders appeared as early as 1950 and inkjet typewriters in the 1960s. In the 1970s virtually every major printer manufacturer invested in inkjet development in an attempt to replace impact (matrix) technology. Reliability and print quality have long been issues because it is difficult to control the ink flow and to prevent the ink from drying and clogging the printhead. The print quality depends heavily on the complex relationship between ink, printhead, and receiver material. By the end of the 1980s, Canon and Hewlett Packard had mastered both the ink chemistry and the hydrodynamics required. In 1998, the total narrow format coated inkjet media market was good for approximately 360 million square meters. Today it is ten times that.

Liquid inkjet printers generally fall into two classes: continuous and drop-on-demand. In a continuous inkjet printer, a continuous spray of ink droplets is produced and the unneeded droplets are deflected before they reach the paper. Continuous inkjet technology provides high-speed drop generation at 1 million drops per second or faster. Two classes of continuous inkjet products are available:

- High-speed industrial printers for carton and product marking, addressing, on-demand short-run printing, and personalized direct mail.
- Proofing printers for verification of materials prior to printing on printing presses. These printers offer the best print quality of any non-photographic device but are much slower (less than an inch per second). Resolutions are not high (e.g., 300 dpi) but variable-sized dots make photographic quality possible.

Drop-on-demand inkjet printers produce ink droplets when needed. The two technologies to drive the droplets out of the printhead are thermal (used by HP, Lexmark, Canon, Olivetti, Océ, and others) and piezoelectric (used by Epson). Thermal printers have been successful because they are inexpensive. The biggest challenge for piezoelectric inkjet technology is the cost and difficulty of producing printheads. Today Epson has a very successful line of piezoelectric color printers, namely, the Stylus color and Stylus photo family of printers offering photographic quality.

3.2.1 Color Laser Printers

In 1938, Chester Carlson, a patent attorney and a graduate of Caltech, discovered a dry printing process called electrophotography. For 9 years, Carlson tried to sell his idea to more than 20 companies including RCA, Remington Rand, General Electric, Eastman Kodak, and IBM. They all turned him down, wondering why anyone would need a machine to do something you could do with carbon paper. Enter Haloid, which became Xerox and the rest is history.

In 1969, Gary Starkweather, at Xerox's research facility in Webster, New York, used a laser beam with the xerography process to create a laser printer. In 1979, IBM introduced the IBM 3800 laser printer, capable of printing 20,000 lines per minute.

In 1978, Xerox introduced the 9700 laser printer. This was the first laser printer commercially available in the United States and in the world. It could output 120 pages per minute. It is still the fastest commercial laser printer. However, the 9700 was physically too large and carried a large price tag as well.

In 1984, Canon launched the LBP-CX laser beam printer. Having many unique advantages over other processes, xerography was adopted for computer output printing. Xerox was the forerunner in this endeavor.

In 1984, HP marketed the LaserJet printer (8 pages per minute). The major feature of this printer was its use of an operator-replaceable all-in-one toner cartridge. The entire development subsystem was built into this toner cartridge. Canon had used this concept in a line of desktop copiers.

The first desktop color laser printer was the QMS Colorscript 1000 in 1993, selling for $10,000 and based on a Hitachi engine using a dual-component toner design. Only a few thousand machines were sold. Annual shipment volumes of color lasers have steadily increased since then. In 1995, QMS cut the price of the Magicolor LX to $4995. By late 1997, QMS was offering the PostScript-enabled Magicolor 2 CX for $3500 and, by early 1998, the Windows-only Deskla-ser 2 for $2500.

Two engines dominated the installed base of color lasers: the Konica engine in the original HP Color LaserJet and LaserJet 5, and the Canon EP-H engine used in many models, including the Lexmark Optra C. In 1997, QMS introduced the Magicolor 2 line based on a Hitachi engine and Minolta brought a new engine to market. The Minolta Color Pageworks saw models based on this engine priced under $2000. In 1998, Tektronix introduced the $1800 Phaser 740 based on a Matsushita (Panasonic) engine. Canon introduced two new engines sold by HP as the Color LaserJet 4500 and 8500.

Pricing for color printers has fallen as volume drives manufacturing costs and as shipment volumes increase; cost is lowered and the end-user price is lowered, which drives shipment volumes even higher. Cost per page has constantly dropped as original equipment manufacturers (OEMs) adjust the pricing of their supplies so that color is finally affordable for corporate buyers. Monochrome printing costs for color lasers are an issue. The cost of color is only a little more than the cost for monochrome, and OEMs are positioning color lasers as economical as monochrome-only printers.

Quality continues to improve and speeds increase with each generation of engines. Color laser generations are shorter than monochrome generations. This is driven by a combination of declining printer pricing, declining costs per page, higher print speeds, and improved quality. These rapid advancements make engines obsolete more quickly than in the monochrome market. Every OEM desires to establish its brand in the color laser printer market, and no one has achieved in color laser printers what HP/Canon achieved in monochrome laser printers. The color laser market is a chance to command a significant market share of machine shipments that will drive long-term profitability in consumable supplies. HP probably will not dominate the color laser printer market to the level it did in the monochrome market. From the late 1980s to early 1990s, HP held more than 80% of the market share in monochrome lasers. In 2000, HP held about 50% of the market share in color laser printers.

The value of supplies for color printers is significantly higher than monochrome, and OEMs understand that the path to higher profits lies in increased market share. Driving this move toward color lasers is the closing gap between the pricing of monochrome-only printers and color printers. As the price and speed of color lasers become comparable to monochrome lasers, users perceive that they are getting color for free if they buy the color laser.

Because the color laser prints both monochrome and color documents, essentially doing the work of two printers, when the purchase prices of color printers drop to about 150% of the cost of comparable monochrome systems, you get color for free. The market for letter/legal size–format color lasers has broken through this 1.5× price point level for several models. Monochrome lasers (letter/legal size) with print speeds of 16 ppm to 20 ppm range in pricing from $1000 to $1500. Pricing levels at 1.5 times this range would be $1500 to $2250. Entry-level color lasers offering monochrome print speeds from 12 ppm to 16 ppm are priced from a low of $1300, with many models in the $1500 to $2000 range, and HP's Color LaserJet 4500 sells for $2200 to $2300. Color is becoming a viable option for multiple business scenarios.

In 2002 prognosticators consistently overestimated how quickly the corporate market would adapt color lasers. Annual U.S. shipment volumes for color lasers range from low-side projections of 800,000 units to high-side projections of 1,800,000 units for 2002 and 3 million units in 2005. The large range in future shipment projections is due to uncertainty in the speed of color adaptation by businesses. If the color laser market does meet these high-side projections, the overall installed base of color lasers will explode by a factor of ten over the next 3 years compared to the current installed base. There are more than 20 desktop color laser engines in the market, and additional engines are being released as this market accelerates. The gap between HP/Canon and the rest of the pack is much smaller in the color laser printer market than it was in the early years of the monochrome laser.

Color lasers use the same fundamental electrophotographic steps as monochromes. The primary difference between monochrome and color is how to perform this process four times for a single sheet of paper, once for each of the four colors of cyan, magenta, yellow, and black (CMYK). All monochromes directly transfer the toner from the organic photoconductor (OPC) device (drum or belt) to the paper. Color lasers use this direct transfer design and some also utilize an intermediate transfer mechanism.

Another variation in color laser printer design is in how the four process colors are positioned to develop toner to the OPC device. Some use fixed positions for the four toners, whereas other designs rotate the colors on a carousel device. The majority of color lasers build a complete four-color image either on the OPC device or the intermediate transfer device. It is then transferred to the paper in a single operation or pass. An exception is the Optra C (Canon P320/ EPH) design, which transfers each of the four colors in succession to the paper.

The toner composition of most of the color lasers introduced since 1994 have been a monocomponent, non-magnetic design. Image development is accomplished via electrostatics. The Konica (HP Color LaserJet), introduced in 1994, was a dual-component design. A recent development has been the use of monocomponent, magnetic black toner in combination with monocomponent nonmagnetic cyan, yellow, and magenta toners, first introduced in the HP 4500.

In color laser printers, there are at least eight replacement supplies: four toner units and an OPC unit, plus a transfer unit, fuser oil, fuser cleaning units, maintenance kits, multiple charging units, four developer units, and other engine-specific replacement units. Monocomponent toner designs reduce user replaceable units.

3.2.2 THERMAL WAX TRANSFER

Since they were introduced by QMS in 1988, thermal wax printers have captured a significant share of the low-end color output market, in part because of falling prices. The first such devices were in the $30,000 range, whereas recent thermal wax printers are less than $2,000. Among the leading vendors of thermal wax printers are QMS and Tektronix.

Thermal wax transfer is a printing process that transfers a waxlike ink onto paper. A Mylar ribbon is used that contains several hundred repeating sets of full pages of black, cyan, magenta, and yellow ink. The ink applied in the thermal wax transfer methods is solid at normal temperature. A sheet of paper is pressed against each color and passed by a line of heating elements that transfers the spots, or pixels, of ink onto the paper. The back of the color ribbon is in   direct contact with the heated surface of the thermal head, which reaches temperatures of 350°C. The image is built up of the three subtractive primary colors, cyan, magenta, and yellow, with black as an optional extra. The colors are carried on the ribbon as sequential panels, and in the printing process, the ribbon moves forward continuously while the receiver is recycled underneath it, so that the whole image is written in yellow, then in magenta, and finally in cyan. The principle of the thermal transfer is:

1. Send print paper together with an ink ribbon into the space between a head and a platen roller.
2. Run electric current into the thermal head to heat it up.
3. The head and roller put pressure on the print paper and ink ribbon as they go through.
4. The heated thermal head fuses the solid ink applied on the ribbon into liquid.
5. The liquid ink is transferred from the ink ribbon to the paper and absorbed by the paper.

The amount of ink absorbed by the paper is controlled by the amount of electric current running into the thermal head. By controlling the amount, gradation reproduction by ink density is possible in this printing method. The major advantages of thermal wax printers include their relatively low cost and the high opacity of the wax, which makes them ideal for creating overhead transparencies. The disadvantages of some thermal wax printers include speed and quality, plus the need for special paper.

These color printers were used primarily for proofing design and artwork in the graphic arts market.

3.2.3 Inkjet Printing

The principles of inkjet printing have been known for hundreds of years, and inkjet devices have been constructed for more than 100 years. However, the technology has only been applied commercially since 1970. Since then, date-coding requirements and the move toward a databased society have driven development in this area. Inkjet printing is a form of non-impact printing. The first inkjets were created in the 1970s by Dr. C. Hellmuth Hertz, a physics researcher at the Lund Institute in Sweden. Inkjet printers have become increasingly essential in the wake of desktop publishing because of the great demand for the high-quality printers for text printing and full color printing. There are two kinds of inkjet printing: drop-on-demand (DOD) and continuous. Continuous inkjet printers have an advantage over DOD printers because of their ability to run at high speed; DOD printers produce high-quality images that closely resemble those of a photograph.

Inkjet printing has a distinction that sets it apart from most other printing technologies: it is a plateless process. The general principle behind inkjet is that the ink is sprayed onto the paper. The device that sprays the ink onto the page is generally referred to as a head. Inkjet systems can be expanded (increasing the size of the paper or their speed) by increasing the number of heads they print with. The way the ink is sprayed onto the paper is generally how the different types of inkjet systems are classified. One approach is to use a constant stream of ink from the jet known as continuous systems. These systems use water-based inks.

A second approach is to send ink through the jet only if a dot is meant to be on the page; these systems are known as DOD. Within the DOD family, the systems can be divided further into what type of inks they use. Most DOD systems use water-based inks, but some use inks that are solid at room temperature. These systems are referred to as phase-change printers because the ink starts as a solid, is melted, and then solidifies on the page. Continuous inkjet, as the name would suggest, uses a continuous flow of ink. With this constant stream of droplets, the system simply has to control whether or not they hit the page. This task is accomplished with the use of electric forces. As the droplet leaves the orifice, a charge is placed on it. The drop then travels between two oppositely charged plates. The charged drop is attracted to one plate and repelled from the other and its path is controlled. The drop either travels to the page or to a gutter that returns it to the ink reservoir. An advantage to this system is that the continuous flow does not allow ink to dry and clog the jet.

DOD printing devices control whether or not the ink hits the page by controlling whether a droplet is formed or not. The ink is contained in a reservoir and a force is required to push it out onto the page. Some systems use heat and others use the deformation of piezoelectric crystals. In systems that use heat to generate the droplet, the ink supply is heated until a portion of the ink vaporizes and this expansion within the reservoir forces a droplet out of the reservoir onto the page. In systems that employ a piezoelectric crystal, an electric current is run through the crystal, which causes it to change shape. This change of shape in the crystal forces ink from the jet onto the page.

Another form of DOD inkjet printing is known as phase change inkjet. In phase change inkjet the ink starts out as a waxy solid, is melted by the printhead, and returns to a solid state once it reaches the page. The system used to spray the ink onto the page is similar to the DOD process; the only difference is the characteristics of the ink. In phase change inkjet, the fact that the ink becomes a solid when it hits the page gives the process some unique advantages. This ink can be used with a very broad range of substrates, because it bonds with the surface instead of depending on absorption. The colors of these inks may be vibrant because of their position on the paper.

There is a tremendous variety of inkjet printers available. This variety includes systems with different qualities, speeds, and prices. The simplicity of the principles behind inkjet is what leads to its many different forms. Basically, all inkjet does is spray ink onto paper. There are almost limitless ways to do this. The ink, paper, number of heads, size of jets, type of ink, and configuration of the printer are only a few of the options that have already been explored. An example of some of the variations can be seen in the continuous process. Some continuous inkjet printers use large droplets, producing low-resolution results but at a very fast rate, and others use very tiny jets and produce photo-quality images. The configuration of the printer may also be specialized. The paper, heads, or both may move as the image is written. Some printers are only large enough to address an envelope; others are large enough to print a billboard. The freedom from having a fixed-image carrier and the ability to be configured in so many ways are inkjet's greatest assets.

Inkjet technology encompasses a wide spectrum of systems — at one end of the market is a Canon desktop color printer that challenges Epson's position as the quality inkjet supplier at under $500. At the other end of the market, there is the Scitex Digital Printing's (now Eastman Kodak) VersaMark press for on-demand book printing at 3,800 book pages per minute at more than $1 million. Also consider Agfa's agreement with Xaar to develop Xaar's page-wide inkjet printheads for potential use in future Agfa digital color presses, and it appears that inkjet is on the move. Inkjet technology is an established technology for much of the digital contract color proofing market and is found in products such as the Creo Iris (now Kodak Veris) proofers and the DuPont Digital Cromalin.

The concept proofing markets, with products from Canon, Epson, Hewlett Packard and Lexmark that sell for less than $2000, are likely to take over the proofing role currently filled by dye-sublimation products and the contract proofing inkjets mentioned above. The quality of these devices is exceptional for their price and getting better.

Wide-format inkjet units from companies such as Encad, Hewlett Packard, ColorSpan, Raster Graphics, Roland, and Xerox are changing the point-of-sale and sales display markets, and impacting technologies such as screen printing. They are also providing full-color imposition proofs to help speed the acceptance of computer-to-plate in the marketplace.

For the very highest speed, there is continuous inkjet, which uses an array of heads, as in the Scitex VersaMark system. This technology is just now reaching 600 dpi. The technology for the future may be DOD inkjet using piezo or thermal technology. This is already operating at resolutions of up to 1440 dpi, with individual drop sizes comparable to that of film imagesetters. In the low-cost desktop printers, these units have heads with a limited number of inkjet nozzles, and the head is moved across the paper to print a line. This gives high quality but restricts the speed. In desktop color proofing, this is not a problem. In applications using technology from Agfa and Xaar, speed and quality are required, and page-width printheads of either 9 inches or 12.6 inches, each with thousands of nozzles, are necessary.

For such applications, it is anticipated that there will be four printheads, one for each process color, printing in a continuous mode on one side of a sheet of paper. Using a page-width printhead allows a significantly greater number of droplets of ink to be generated per second. The potential is to develop a far simpler 100-page-per-minute-plus digital color press than can be achieved using electrographic toner-based technology.

Piezo inkjet technology is a very simple process compared with the complexity of continuous inkjet technology. The difference is speed. Large-format inkjet printers will become the standard proofing devices for printers. A number of large-format engines and desktop devices use similar technology (such as the new Epson 9000), and large-format devices could also be used for contract-quality color proofing. Very-large-format printers from Matan, Idanit, and others for

billboard and poster printing will impact lithographic- and screen-based technologies. For high-speed digital color printing, inkjet will challenge the position of
the electrophotographic systems such as those from Indigo and Xeikon and
forthcoming systems from Xerox, NexPress, and others.

Canon's bubble jet, desktop color printer challenges Epson's position as the
dominant printer vendor in the low-cost color proofer market. The A-3 size, 1200-
dpi printer called the Aspen first appeared more than a year ago when it was
previewed as a concept printer at Comdex. It appears to break through in printing
technology with its six colors plus coating capability. The BJC-8500 uses Canon's
bubble jet inkjet technology; Epson uses piezoelectric inkjet technology. The
printer also features a Canon-trademarked technology called MicroFine Droplet
Technology. This technology includes a new printhead design and enhanced ink
formulations that make a microscopic drop size possible at the unit's maximum
output resolution of 1200×1200 dpi. This technology ensures that each drop not
only is consistent in size but also is positioned with precision accuracy and high
velocity.

The BJC-8500 does not lose printing speed to achieve this quality. The new
printhead has 1536 nozzles (which include among others 256 for all colors, 512
for black, and 256 Ink Optimizer nozzles). The volume of nozzles allows the
head in a single pass to print two or three lines of text, depending on the type
size. The printhead is made up of two modules and works in two print modes.
In CMYK mode, one cartridge has the cyan, magenta, and yellow capability and
the other houses the black and print optimizer.

In photo-imaging mode, the black and print optimizer cartridge is removed
and a second three-color cartridge is installed (light cyan and magenta, plus
black). In every case, all cartridges are fed by separate independently installed
ink tanks, one for each color. Individual ink tanks can be replaced as they run
out, instead of replacing an entire cartridge.

A key feature of the BJC-8500 printer is Canon's P-POP (Plain-Paper Optimized Printing) system. This treats the print surface of the paper with a clear
liquid solvent, Ink Optimizer, just milliseconds before the ink is ejected from the
printhead. The feature, in effect, transforms the plain paper into coated paper for
water resistance, sharper detail, and more vibrant colors. If the unit is printing
onto a coated stock, then the six-color photo-imaging mode will give a wider
gamut for printing without the need for the Ink Optimizer. The sheet feeder of
the unit handles stocks up to a maximum of 13×19 inches, allowing an oversize
A3 page to be printed with full bleed.

The result on a high-gloss stock using six-color printing is outstanding. The
1200×1200 dpi resolution and MicroFine Droplet Technology provide very fine
control, probably greater than Epson's 1440×720 resolution, which has a larger
droplet size.

Inkjet printers have a wide range of applications in the printing and packaging
industries. Inkjet printers can be used for marking products with dates, such as
"best before," as well as coding information such as prices and product tracking.
There are hundreds of different products that can be coded using inkjet printers:

food and beverage containers, cosmetics, pharmaceuticals, electronic compo-nents, cables, wiring, PVC and P. E. pipes, glass and plastic bottles, and industrial components. Ticket numbering and high-speed addressing for magazines and direct mail are just two of the applications in the printing industry for high-speed inkjet printing.

The Hertz technology that created inkjet printing is known for the high-quality images it can create. The main reason that the quality is so high is that the inkjet can produce true halftones, such as different gray levels, or color tones, which can be generated with every single pixel. This true halftone printing is achieved by varying the number of drops in each pixel. The number of drops may vary from 0 to about 30 for each color, which means that a number of different density levels for each pixel and color may be obtained. It is possible to increase the number of density levels per pixel from 0 to 200 for each color.

3.2.3.1 Continuous Inkjet Printing

Continuous printing speed is approximately 45 square inches per minute or even higher. Since the introduction of the continuous inkjet printer, there have been several modifications and improvements to the system. The two main concerns of the early inkjet printers, nozzle clogging and uncontrolled ink mist, have been addressed by the creation of the IRIS continuous inkjet printer and others.

After each cycle is complete and the nozzle has stopped firing, the nozzle tips are vacuumed to remove any residue ink. An automatic nozzle maintenance cycle is built into each system. When the printer is not in the print mode, the system powers up on a timed cycle and fires ink through the nozzles for a few seconds, shuts down, and vacuums the tips again.

Uncontrolled ink mist is a result of the reaction between the dropping ink and the printing surface. Ink droplets are forced out of the nozzle at about 650 pounds per square inch of pressure. This means that the droplets travel at about 30 mm from the nozzle tip to the print surface at a speed of 20 meters per second, or 50 miles per hour. The mist develops from the millions of drops that are hitting the paper every minute. A mist shield was created to control random ink spots.

The mist shield consists of an absorbent material positioned near the substrate printing surface that catches the ink as it bounces back toward the ink nozzle. This allows for a clean print surface with fewer random background spots as well as clean internal surfaces.

In the IRIS system, there is a print resolution of 240×240 dpi with lateral printing speeds of about 1 or 2 inches per minute. The IRIS 2044 is a large-format system, and images covering the 34- $\times$ 44-inch maximum size can be created in about 30 minutes. There is also an IRIS 2024, a medium format that can produce images covering a maximum of 24×24 inches in a printing time of 15 minutes. The substrate is manually sheetfed so that the printing sequence can always be varied.

The quality of the color image produced on an IRIS system, in terms of the resolution and the color shades that are possible, depends on the number of separate color dots per square inch that compose the final page. This technology uses more than 230,000 dots per square inch to compose a full-color image at 240 dpi per color.

Continuous inkjet printers are used for a variety of useful applications, including bar coding, pharmaceuticals, and food packaging. They have a wide range of specifications that make them very flexible for printing on various substrates of all shapes and sizes. Bar coding seems to be one of the most popular applications of continuous inkjet printers. They can produce medium-density codes and are able to print up to 17 mm high, which means that the bar codes produced may only be read in what is called a closed system.

Closed systems are controlled systems in which reading or scanning devices are necessary, such as the scanners found in grocery stores. Bar-coded products can be scanned by these systems. If alphanumerics are going to be printed under a bar code, the option exists of using an inkjet with two printheads: one to print the bar code and one to print the alphanumerics. In general, inkjet printing is used to print on the product directly. The unique construction of the jet valve allows for printing on non-absorbent materials such as PVC tubing, shrink films, or ceramic tiles.

Scitex Digital Printing has a digital printing system called VersaMark for books capable of 3800 book pages per minute at 300×600 dpi. It uses continuous inkjet on a web printer at three times the resolution of the 240-dpi earlier models. The speed is achieved by printing three 6- $\times$ 9-inch pages side by side on a 20-inch web. For 8.5- $\times$ 11-inch pages, the rated speed is 2100 ppm.

There are two 9-inch imaging heads, each with 2600 inkjet nozzles for monochrome printing, but the system is also applicable for spot-color applications. Binding is a separate step. Scitex is not attempting to raster image process (RIP) incoming jobs at the speed of the imager. The RIP supplier is Varis and portable document format (PDF) is the standard format for input to the RIP. The cost per page is less than half a cent, or about $1.30 to print a 300-page, 8.5- $\times$ 11-inch book, one of the least expensive production approaches. The VersaMark is priced at $800,000 to $2 million, based on configuration, with the first customer site in place.

In contrast to continuous inkjet printing, in which the droplet stream is continuously expelled from the nozzle (with the unwanted drops caught and recycled into the ink source), DOD forms and expels droplets only at the moment needed. DOD printers provide a larger letter size, the ability to use more than one printhead, and a less critical viscosity requirement for the ink. Most use multiple nozzles. Each system has limitations because of the inter-relationship between image height, image quality, and print speed, determined by the size of ink drops and the rate of production. There are three subtypes of DOD: piezo-electric liquid, bubble jet/thermal liquid, and solid ink. In piezoelectric crystal inkjet printing, stress on the piezoelectric crystal produces an electric charge that causes the droplet to be expelled. The advantages of this method include reliability and the potential for high speed. The disadvantage is the relatively high cost of manufacturing. These systems are sometimes supplemented with a jet of air to impel the ink drop.

The bubble jet printer is a development of HP and Canon. The first commercially available bubble jet printer was the Hewlett Packard Paint Jet, a letter-size, monochrome printer with 180-dpi resolution. It was capable of printing on coated paper and transparency film. In 1991, HP introduced the DeskJet 300C, the first 300-dpi color printer. A three-color printer, the DeskJet 300 could not print a true black. This was remedied in 1992 with the four-color HP 550. Canon entered the market in 1992 with its line of BubbleJet printers. Today, most bubble jet printers, including those made by HP, Apple, Star, and Lexmark, use the Canon engine. The Epson Stylus line of color printers offers resolutions of up to 1440 dpi.

In bubble jet/thermal liquid inkjet printing, an electric charge is applied to a tiny resistor, which causes a minute quantity of ink to boil and form a bubble. As the bubble expands, a drop is forced out of the inkjet nozzle. The current applied to the resistive heating element causes bubbles to form; small bubbles consolidate, and the pressure begins to expel the drop. The bubble continues to expand and push out the drop; as the drop is expelled, the heating element cools and the pressure of the bubble is countered by the pressure of the ink. The bubble is expelled and the system returns to the wait state for the next bubble.

Advantages of the bubble jet/thermal liquid ink method include low cost, excellent print quality, and low noise. However, liquid ink, for both piezoelectric and bubble jet, generally requires special coated paper, which is expensive and limits the applications of these printing methods.

The third type of inkjet printer uses a solid ink (also called hot-melt). A wax-based solid ink, similar to a crayon, is quickly melted and then jetted to the paper. The ink solidifies on contact, preventing smudging. The advantages of hot-melt include the ability to print on a variety of substrates, excellent print quality, and the potential for high speed. Experts cite advantages of hot-melt as low cost, high reliability, flexibility, media independence, user safety, and environmental friendliness. Color and image quality are independent of substrate properties, and the prints are water resistant and lightfast.

One main disadvantage of hot-melt is clogging of the nozzle by dried ink. Hot-melt inks perform poorly on transparency film, because they scatter light too much. This can be controlled by precise temperature control within the printer, on a platen. Also, because the wax ink sits above the printing substrate, it has reduced resistance to abrasion, cracking, and peeling. Different manufacturers use different methods to compensate for this.

The first solid ink printer was the Phaser III PXi tabloid printer, costing roughly $10,000. Introduced in 1991, it could print with a resolution of 300 dpi on a page up to 12×18. It featured a 24-Mhz reduced instruction set computer controller and used Adobe PostScript Level 2 page description language. It could print full-color pages in 40 to 60 seconds and monochrome pages in just 20 seconds. Dataproducts, owner of several key patents in the solid ink area, offered its Jolt printer in 1991. Brother is also a player in this field, with its HS 1PS.

3.2.3.2 Technological Developments and Considerations

Continuous inkjet was the first form of inkjet printing and still predominated as of 1994. It boasts higher resolution and higher speed than DOD. However, over the past few years, developments in DOD technology have given rise to a wide range of new applications, from photo printing to wide format printing to package printing. The advantages of DOD include low cost, compact size, quiet operation (only the ink strikes the paper), excellent print quality, the ability to produce excellent color, and the potential for high speed.

It is essential to consider the DOD printer as part of a larger system of printer, ink, and substrate. To that end, considerable research into the physics and chemistry of DOD printing has been conducted. The frequency response and print quality of a DOD inkjet system are determined principally by the time taken to replenish lost ink after ejection of a drop. Mathematical analysis was extended to multijet, systems connected to a common reservoir, and it was found that if all jets were fired simultaneously, refill was slow and drops were large, due to the inertia of ink and overfilled tubes. Sequential firing provides quicker refill due to uniform ink flow rate and minimum inertia.

Much attention has been given to the development of ink. Considerations include temperature dependence of ink properties, aging, evaporation loss, acid-base resistance, corrosion, and ink mixing. Also important are viscosity, pH, surface tension, dielectric properties, optical density, and environmental impact. Aqueous inks were the first to be used, but their limitations (mentioned above) led scientists to research both improvements and alternatives. Clogging has been another sticky problem. In DOD thermal inkjet printers, nonvolatile fluids are mixed with the primarily water-based inks to minimize precipitation in the nozzles, which can result from evaporation of the water.

This problem has been addressed by using highly water-soluble, low-molecular-weight additives that are claimed to be 3 to 5 times more efficient than glycols in reducing water evaporation and useful for prevention of nozzle failure in DOD printers. It is believed that the additives form a readily ruptured membrane at the surface of the nozzle during the dormancy period of the jets. Questions of droplet air drag and the effect of electrical charge on the ink have also been studied. For most liquid ink printing, the substrate is still more critical than it is for hot-melt. The transparent inks appear to better advantage if they are not absorbed into the substrate.

Accuracy of reproduction depends on constant diameter, regular contour, maximum intensity, and instantaneous drying. In 1985, the French firm of Aussedat-Rey produced its Impulsion paper, which was claimed to be of the correct absorbency for accurately timed drying.

Many printer manufacturers sell their own lines of coated stocks. Paper manufacturers also have their own lines of inkjet papers. Some printers now claim high resolution on plain paper. It is the paper that gives the color output the look and feel of a photographic paper.

3.2.3.3 Inkjet Papers

Papers that are created specifically for inkjet printers have two elements, the foundation and the coating or treatment. The paper foundations are composed of fiber, sizing, filler, and colorant. The coating or treatment is added to the foundation of the paper to further make it optimized for inkjet usage.

Short paper fibers are the best inkjet papers. The paper fibers affect how smooth the paper is and the overall image quality. Paper smoothness is measured in sheffields; the lower the number, the smoother the paper. With smoother paper you can help maximize the printers output characteristics. If paper has a rough surface, the high-dpi equipment's printing capabilities will be lost. Think of it as a landscape: one dot will be on a hill, another will be on the side of the hill, and a third will be in a valley. Paper fiber can be compared to sandpaper grit. The smaller the grit on the sandpaper, the smoother the surface will be. The same is true with paper fiber: the smaller the paper fiber, the smoother the paper.

With shorter fibers, there will be less ink migration and the ink will not travel as much on the surface of the paper. The ink migrates/bleeds because it is being absorbed by other parts of the paper, i.e., osmosis. The result of having shorter paper fibers is a decreased amount of bleeding. With less ink bleed, the colors will not appear as washed out, because the inks will be in a more concentrated location. Smaller fibers and smoother papers have less air between the fibers. The ink will fill these pockets of air and increase the transparency and reduce the reflection density of the image.

Another aspect of preventing ink migration/bleed is using sizing on the substrate. It is added to prevent water absorption into the paper, prevent ink smudge, and aid in drying. Sizing can be internal like rosin or alum, or it can be applied to the surface like liquid starch. The basic principle behind sizing is that it is hydrophobic so once it has reached the saturation point, it will not permit any more absorption into other parts of the paper. Without sizing, there would be an increased amount of bleeding in the paper's surface. The sizing further inhibits inks from migrating to other parts of the paper.

Fillers affect the physical and optical appearance such as brightness, color, opacity, and the feel and stiffness of the paper. Fillers may be clay, talc, titanium-dioxide, calcium carbonate, or other materials. To make the paper appear brighter and have a whiter color, a filler of titanium dioxide can be used. The opacity of the paper will also determine now much light is reflected or transmitted through the paper. The paper surface should be more opaque so that light will not be transmitted but, rather, reflected off the paper.

The combination of paper brightness, whiteness, and opacity contributes to the color quality of the image as well. With a brighter, whiter, and more opaque paper, you have an increased color gamut, more visual contrast, increased saturation, and increased detail and sharpness.

The colorant is used to give the paper a color or tint, such as publication-based commercial stock. In some cases, such as during proofing, the press condition should be mimicked. If the paper has a yellow tint, that effect with the inkjet paper should be reproduced.

After the foundation of the paper is made, it is then put through a treatment or receives a special coating. Do not confuse sizing with the special coating; sizing is one aspect of the coating inkjet paper receives. The treatment/special coating does help with ink absorption at the location where the ink hits the paper. It helps prevent ink migration and bleeding and adds florescence to the paper, making it appear whiter. The paper under a UV light will have a bluish/purplish glow. An uncoated piece of paper will appear white. Sometimes, however, with a coated paper, the surface becomes slick. This sometimes causes problems with the loading and unloading of the paper, because the feed rollers slip on the paper's surface.

Inkjet papers should have proper porosity, be smooth, and have uniform weight and thickness. Porosity is the product of the combination of paper fiber and filler. The proper porosity will allow the right amount of ink to be absorbed into the paper. If too much ink is absorbed, the paper will have too much water in it and the fibers will expand and cause waviness and feathering at the image edge. The resulting image will not appear sharp and may become oversaturated in the center of solids. If the paper is not porous enough, the ink may not be absorbed enough into the paper and cause mottling. Mottling occurs when the ink does not dry because it is not being absorbed into the paper. The resulting image will not look correct where the solid areas are not as solid as they should be.

The smoothness of the paper is a product of paper fibers and treatment and/or special coatings. The smoother the paper, the more directly light will be reflected. The resulting image will appear sharper and brighter. If the paper is rough, the light being reflected will be more diffused. The resulting image will appear flat and dull.

Uniform weight and thickness will reflect the quality and smoothness of the paper. If the paper has much variation internally and on the surface, the print quality will be poor. This is because the ink placed on the paper will be of inconsistent densities of like colors. This can be seen in printing solids. It is similar to the printing of solids on toner-based printers. The solid appears splotchy and looks as though it has high and low points. The more uniform the weight and thickness of the paper, the more consistent the output will be.

A large selection of inkjet papers is available, with varying paper thicknesses; tints; matte, semi-matte, and glossy surfaces; watercolor; textures; and also available are transparency sheets and cloth sheets. The application will determine what paper to use.

The basis of good inkjet paper comes down to good dot structure. Because fluids form spheres in a free state, when the ink hits the paper a nice circular dot should result. Under ideal conditions, the diameter of the sphere should be the diameter of the dot. Even though the dot is an area and the sphere is a volume, the excess should be absorbed into the paper.

Average paper thickness is 4 to 7 mils, and the diameter of the ink drop coming out of the printhead is approximately 10 microns. This means that an ink drop is less than 1/500 the width of the paper. When the ink hits the paper, a dot forms. This dot should be circular and have minimal bleed, dot gain, and feathering. If these three occur, the image should have nice dots, good highlights, midtones and shadow detail, good tonal range, good saturation, good detail, sharp graphics and text, increased contrast, and superior reproduction when compared to a print down on regular paper (such as copier paper).

A technique used in some high-end inkjet printers is to place a charge on the ink drop and have the paper be oppositely charged. The paper itself does not have a charge, but it is given a charge during printing. It is important to keep in mind that inkjet printers use dyes instead of pigments. Even though pigments have a stronger hue than dyes, pigments are not very soluble in liquids; in return, they are too large to fit through a nozzle or piezo of inkjet printhead. The orifice of the printhead is usually smaller than 10 microns. (Iris Graphics, later Creo and now part of Kodak, pioneered the under-10 micron orifice.)

Canon has come up with an interesting concept with its inkjet printers using copier paper. Instead of having the standard four CMYK ink reservoirs, Canon has five reservoirs. The additional reservoir is a liquid that jets before the ink and acts as a special coating. It essentially traps the ink and prevents ink migration.

3.2.3.4 Desktop Inkjet Gets Down to Business

The most frequently cited uses for DOD printing have been low-end office printers and case/carton marking. Other applications include direct mail personalization, labeling, ticketing, bar coding, dating, and marking.

Digital photography and digital printing were made for each other. Whether one uses a desktop inkjet printer for outputting the equivalent of photographic prints or a high-end toner-based printer for document reproduction, digital photography allows for image inclusion more easily than ever before.

The inkjet color printer has become the major printout device for the average user.

3.3 THE PRINTER MARKET

3.3.1 MARKET SIZING

HP has a large global installed base of printer users. The company's first LaserJet and inkjet printers were introduced in 1984, and since then it has shipped 230 million printers, with 130 million inkjet printers and 60 million LaserJet printers. HP is adding 30 million inkjet printers and 7 million LaserJet printers each year. The company has predicted more convergence at the high end of the imaging and printing markets, where enterprise customers will be buying more multifunction products.

HP inkjet printers make up about 52% of the market, IDC estimates, whereas Xerox accounts for less than 4%.

Monochrome page printer sales have declined 5% to fewer than 3 million units shipped yearly, but some segments are growing. Volume growth of all U.S. electronic printers was flat in 2001, but color laser printers and MFPs are strong growth segments.

The worldwide printer market declined by 5.3% in the first quarter of 2002. A total of 18.9 million units were shipped, representing about $5.4 billion in end-user spending.

Gartner Dataquest reported that U.S. color page printer hardware shipments grew 24.6% in 2002, to more than 315,000 units, up from 252,000 units in 2001. U.S. color copier hardware shipments reached 44,000 units in 2003, an increase of 3.5% over 2001 shipments of more than 42,000.

Peter Grant, principal analyst for Gartner Dataquest's Digital Documents and Imaging Worldwide Group stated at a Seybold Symposium: "New color page printer products will span the range from entry-level color lasers for the small office/home office to higher-speed color LED printers that will bring affordable color into the workgroup and enterprise. Vendors competing in the low end color printer segment need to balance their low hardware price with an acceptable supplies price. Users are aware of the high cost to print color and to reach the magic cost per page for color that removes the barrier to crossover from monochrome to color. The cost for a color page needs to equal that of a monochrome with the same toner coverage."[1] His prediction was correct.

In 2002, copier vendors moved their products to higher speeds, where they will not sell large numbers of units but, rather, hope to capture considerable page volumes. Gartner Dataquest analysts forecast significant price erosion for color laser copiers at the low end to maintain market share against color laser printer vendors.

Gartner Dataquest analysts have said that color copier vendors must respond more quickly to color desktop printers that will compete aggressively for office output. Organizations will have multiple options to buy, lease, rent, or outsource and buy only the pages they print.

Color inkjet printers now make sense as laser replacements for most users. Inkjet color printers can reproduce photography of stunning quality. The Epson 2000P was the first desktop printer capable of reproducing photographs of archival quality; Epson claims photograph quality of up to 100 years.

With its acquisition of Compaq, HP saw a break in its printer relationship with Dell. Dell's is not a profitable one; therefore, current strategy is merely reselling HP and Epson printers — with no ink to follow. Dell will make its own printers. Most printers are sold at a loss, and it takes a long time to penetrate an installed base as large as HP's, but Dell will be able to bundle PCs and printers.

In addition to meeting the changing needs of large businesses, paper suppliers are increasingly turning their attention to the SoHo sector. Although this market is relatively small (about 2 to 3% of the total market) at the moment, predictions of potential growth vary widely from 5 to 20% of the overall market in 2005. Optimistic projections state that SoHo users will represent over 40% of the total cut size market by 2007.

3.3.2 CONDITIONS IN MOTION

The U.S. printer industry is going through a transition as inkjet printer shipments were on the decline in 2004, whereas all-in-one (AIO) multifunction product shipments were on the rise, according to Dataquest, Inc., a unit of Gartner, Inc. Gartner Dataquest has said that U.S. inkjet printer shipments decreased 5.8% in the first quarter of 2002, whereas AIO multifunction product shipments increased 98%. Total inkjet printer shipments reached 4.1 million units in the first quarter of 2002, whereas AIO (both sheet-fed and flatbed multifunction products) shipments totaled 1.3 million units.

The color laser copier and color laser printer markets suffered from the softer economy in 2001, but both markets show growth potential if vendors deliver the proper mix. This has proven true and growth from 2002 to 2005 has been robust.

Users are aware of the high cost to print color. Gartner Dataquest analysts have said that offering a standard cartridge and a high-capacity cartridge has worked well for some vendors in helping users over the price hurdle.

At a Seybold Symposium in 2002, Peter Grant said, "More important is to reach the magic cost per page for color that removes the barrier to cross over from monochrome to color. The cost for a color page needs to equal that of a monochrome with the same toner coverage."

3.3.3 FORECAST

Inkjet and toner-based (laser) printers will compete. For the next 5 years inkjet printers will own the low end (high unit placement) of the market and lasers will own the high end (lower unit placement). However, in terms of volume, both markets may be equal.

The fight will be in the 10- to 20-ppm market, where inkjet and laser printers overlap. The main drivers will be photographic output and distributed documents (publications sent as files and then wholly or partially printed out remotely). This will lead to higher levels of quality — which means more than four inks — and higher speeds.

Printers are sold through two main approaches: retail and mail order and bundled with PCs. We do not expect any major changes in channels but there will be fierce competition in terms of price and performance.

REFERENCES

1. Romano, F. *Machine Writing and Typesetting*. GAMA, Salem, NH, 1978.

Part II

Platforms

4 Inkjet

*Ross R. Allen, Gary Dispoto, Eric Hanson,
John D. Meyer, and Nathan Moroney*

CONTENTS

4.1 HISTORY AND INTRODUCTION

It is traditional for historical discussions of inkjet technology to begin with references to the 19th-century studies made by Lord Rayleigh on the stability of liquid jets. This work formed the practical basis for the development of continuous inkjet printers that have found broad application in industrial marking processes and the graphic arts. These devices make use of the Rayleigh instability, which causes a continuous liquid jet to break up into a stream of droplets, which then are electrostatically deflected between trajectories allowing them to print or to collect for recirculation.

Desktop color inkjet printers are all drop-on-demand devices: a droplet of ink is ejected only when a pixel or portion of a pixel is to be printed. Given this chapter is concerned with desktop color printers, a reference to Rayleigh might seem out of place. However, it is worthy to note that Rayleigh studied the dynamics of bubble growth and collapse, and in the process discovered cavitation, the damage mechanism associated with bubble collapse. All inkjet printers must contend with unwanted gas bubbles in the ink supply. In thermal inkjet, a bubble of superheated ink vapor drives the ink droplet out of the nozzle. The subsequent collapse of that bubble on the heater can produce cavitation damage, which, left uncontrolled, can destroy the heater after only a few hundred drop ejection cycles. The successful development of thermal inkjet drop generators lasting hundreds of millions of cycles is based on fluidic and materials solutions to control cavitation. Rayleigh's bubble model was the starting point for computer simulations developed to understand and control the growth and collapse of vapor bubbles. It is appropriate, therefore, to begin this chapter with an acknowledgment of Lord Rayleigh's contributions to inkjet technology in all its forms.

Desktop color inkjet printers first appeared in the early 1980s. These printers used piezoelectric (piezo) transducers to generate ink droplets, had only a few nozzles, and offered resolutions less than 240 dots per inch. Recognizing the future potential of low-cost color printing, a number of companies engaged in early experiments with color images using these devices. Affordable desktop color scanners did not appear until the 1990s, and this meant that access to color digital image data was mainly through the graphic arts. In these early days of color desktop printing, the scarcity of digital color images meant that many people used and published results based on a set of common ones, such as the well-known photo called "The Mandrill." The occasional sight of this baboon today brings back memories to the founders of desktop color printing.

The results were promising, but ahead of their time: the supporting infrastructure of powerful desktop computers, image-capable application software, desktop color scanners, digital cameras, and the Internet were not yet available. Market opportunities for color printers were limited, and many industry observers stated that desktop color printing would not succeed until color copiers became commonly available in the office. In the 1980s, this led to the much-debated but never-proclaimed "Year of Color," when color printing was finally accepted in the office. The first year of color passed unheralded in the early 1990s.

Today, desktop color is so commonplace that inkjet printers compete in a commodity market, and this environment demands that all printers offer photographic-quality color as well as high-speed, high-quality black text on plain paper.

4.1.1 Major Technologies

Two technologies for producing drops-on-demand have dominated desktop color printing since the mid-1980s: thermal inkjet and piezo inkjet. An overwhelming number of inkjet printers use liquid, water-based inks. Solid or hot-melt ink is used in some printers found in graphic arts and office applications.

Thermal inkjet, also called bubble jet by Canon, rapidly heats the ink to produce a tiny bubble of superheated ink vapor. The expansion of this bubble ejects an ink droplet, and its collapse refills the ink chamber. Thus, the only moving part in a thermal inkjet printhead is the ink itself. Hewlett Packard (HP) introduced the first printer based on this principle in 1984, called ThinkJet. ThinkJet had 12 nozzles, each producing up to 1200 drops per second, and used special paper.

Thermal inkjet devices are fabricated from thin metallic and dielectric films on silicon using photolithographic techniques developed in integrated circuit manufacture. This enables many drop generators and their control electronics to be integrated on a printhead. HP's ThinkJet print cartridge is shown in Figure 4.1. ThinkJet's printhead and the electrical interconnect were very compact, and

FIGURE 4.1 HP's ThinkJet print cartridge.

ThinkJet offered the innovation of a disposable printhead and ink supply. Perhaps more than any other feature, this concept legitimized inkjet by overcoming the reliability and (messy) ink-handling issues that gave earlier inkjet products a poor reputation. Thermal inkjet technology has undergone significant development primarily by HP, Canon, and Lexmark.

Piezo inkjet technology benefited from substantial investment, development, and innovation by Epson, who introduced the Stylus 800 office text printer in 1993. The technology behind the Stylus printer is a push-mode, multi-layer, piezo actuator. This made possible nozzle generators of compact design, although not small enough to match the high linear nozzle density of thermal inkjet. Piezo inkjet also benefited from photolithographic fabrication of its silicon and metal layers, and this enabled the mass production of printheads at modest cost.

Solid inks have been investigated in both thermal and piezo inkjets, but only piezo solid inkjets have seen systematic product development by Tektronics (now Xerox), Brother, and others. The Tektronics Phaser III, introduced in 1991, had 96 nozzles and operated at 8 KHz. By convention, operating frequency is the number of drops generated per second from each nozzle and is expressed in hertz (Hz). After a decade of technology development, the Phaser 8200 offered a printhead with 448 nozzles operating at 36 KHz. These printers have focused on business and graphics markets offering A-size (A4) and B-size (A3) formats. Although much larger than their liquid ink counterparts, solid inkjet may be placed on a tabletop in office environments.

4.1.2 YEARS OF COLOR

The introduction of HP's PaintJet Printer in 1987 was the logical precursor to the explosive technological development of desktop inkjet color printing. Prior to PaintJet, inkjet products competed to produce black text on plain paper with the goal of meeting the rapidly advancing quality standards set by laser printers. PaintJet offered 180 dots per inch (dpi) color graphics on a special coated paper and overhead transparencies. It used 30 nozzles for its dye-based black ink and 10 each for its cyan, magenta, and yellow inks. The black and color printheads were integrated into two user-replaceable print cartridges.

Color displays for desktop computers were in common use by the time PaintJet was introduced, and the issue of matching the display to the printed image quickly became a key user need. This coincided with the beginning of industry-wide adoption of color management, which drove standards for color space definitions, file formats, and color data conversion methods making possible device-independent color across display, capture, and hardcopy devices from different manufacturers.

In 1989, the HP DeskJet 500C offered 300-dpi resolution, plain paper color, and a printer driver that provided color matching between display and print. Market response to the DeskJet 500C was dramatic, and in the decade of the 1990s, HP produced and delivered more than 100 million color inkjet printers worldwide. The success of desktop color inkjet printing involved much more than higher resolution, smaller drops, and faster throughput. The practical adoption of the printing technology was made possible by research in color science, image processing, color data interchange standards, and the availability of user-friendly color management software.

The 1990s saw the quality of images on special media increase dramatically with every product introduction cycle, roughly one every 6 months. These were the years of the "dpi wars," in which some manufacturers claimed the highest possible resolution in photo printers often at the expense of throughput and the ability to print high-quality black text on plain paper. The introduction of the HP PhotoSmart printer in 1997 with six photo inks demonstrated that a desktop inkjet printer could achieve the goal of photographic print quality.

By 2000, desktop color inkjet printers had created a digital imaging revolution by printing photos on special media that were indistinguishable from conventional photographic prints from 35-millimeter (mm) film. Developments quickly followed in consumer digital photography and desktop color scanners to make high-quality digital capture pervasive and affordable. For example, after 2001 consumers purchased more new digital cameras than new 35 mm cameras (excluding one-time use 35 mm cameras).

Today, the inkjet printing process is no longer the limiting factor that determines the grain, color quality, and color fidelity of the printed image, it is the process of capturing the digital image with either a digital camera or a color scanner. For scanners, the quality of the original image is often the limiting factor.

4.1.3   ADVANCEMENTS IN THERMAL AND PIEZO INKJET

The reproduction of high-quality color images requires sufficient dynamic range along with precise control of tone reproduction, neutral gray balance, image granularity, color gamut, detail, and sharpness. These needs have been addressed by technologies allowing higher resolution and addressability, decreases in dot size, multi-level printing with dark and light primaries including grays, and halftoning techniques. Fade resistance, driven by the durability of conventional color photographs, has been achieved by cooperative design of dyes and special inkjet papers and by the introduction of black and color pigment inks. Today, color inkjet prints can resist fade for more than 100 years and exceed the fade resistance of color photographs produced by silver halide and color-coupling chemistry.

Since 1989, the printing resolution claimed by liquid inkjet manufacturers has increased from 300 to 5760 dpi. This rapid development was driven by a highly competitive marketing situation in which dpi was promoted as a primary, and sometimes the only, image quality specification.

A significant amount of confusion and misinformation surrounds dpi. This is because resolution has been used in a number of different ways since it became a key specification used to guide purchasing decisions. Initially, resolution meant the (horizontal or vertical) spacing of printed dots with sufficient overlap to achieve 100% area coverage. This is what could be called *true resolution*. Resolution has also been used in place of *addressability*, the horizontal and vertical grid at which an ink droplet can be placed. High addressability has value in achieving smooth, sharp-edged characters in text printing. The use of multiple drops per pixel (color layering) or drop volume modulation has made it possible to vary the final dot size while printing at a fraction of true resolution based on the smallest dot. When inkjet printers print photos using halftone pixels offering 72 million addressable colors, as claimed by HP for its 8-ink PhotoREt Pro printers in 2003, it is meaningless to characterize a printer's image quality by a single number such as dpi. Studies based on typical observers conclude that 4- by 6-inch prints produced by a desktop inkjet at 300 pixels/inch using a 6-color ink system are indistinguishable from conventional photographs in terms of image grain, color gamut, smooth color gradation, sharpness, and detail.

As resolution has increased, drop volumes have also decreased, from 32 picoliters (pl) in 1997 to 2–6 pl in 2002. Ink drops are commonly specified by their volume (picoliters, 10^{-12} liters) or weight (nanograms, 10^{-9} grams). As most liquid inkjet inks for the desktop have a high water content, the number of picoliters in a drop is equal to its weight in nanograms within a few percent. This trend up to 2002, shown in Figure 4.2, has been driven by competitive forces to deliver photographic image quality. At 2 pl, drop volumes are approaching a practical limit where aerodynamic forces limit accurate drop placement. These small drops produce dots that are virtually invisible, but used in halftoning and edge-enhancement, algorithms give fine control over the tone reproduction curves, especially in the highlights, and smooth edges on printed characters.

Because smaller drop volumes make smaller printed dots, more drops per second from a printhead are required to cover the same area in a given time. The time to print an image or page is a key competitive specification, and this translates into higher drop rates and more nozzles per printhead. The potential number of drops per second from all the printheads taken together is a useful measure of design capabilities, although such rates cannot be achieved in any practical print mode. By this measure, before 2000 desktop color inkjet printers could not produce more than 20 million drops per second. By mid-2005, an HP Photosmart 8250 printer could deliver over 100 million drops per second from 3900 nozzles integrated onto a single silicon chip. A trend observed over the past 20 years of desktop inkjet printing shows that the potential number of drops per second doubles every 18 months. As practical mechanical and fluidic limits in drop generator design are approached, this trend can continue only by increasing the number of nozzles.

Practically speaking, the throughput of desktop printers is limited now not by drop rates but by the time needed to dry each sheet, and this has driven the development of faster-drying inks and print media. Most desktop inkjet printers do not use heaters to dry the printed sheet because competitive forces in the marketplace cannot support the additional manufacturing costs.

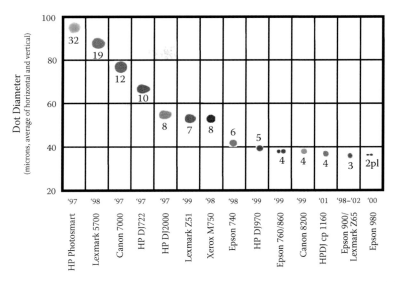

FIGURE 4.2 (See color insert following page 176.) Evolution of drop volume and drop size.

4.1.4 Advances in Inks and Print Media

Alongside developments in printhead technology, a major effort has been made in the development of inkjet inks and print media.

The early days of desktop inkjet were dominated by the goal of laser-quality printing on plain paper, which meant copier and bond (typewriter) papers. A restriction to special inkjet papers, especially to achieve acceptable text quality, was understood to be a significant barrier to the adoption of inkjet products in the office. This drove the development of black inks, especially pigment-based black inks, with high optical density and minimal spread to compete with laser printers on text quality.

As desktop color inkjet printing spread into the office and home, markets developed for special media offering the highest image quality. Today, several major requirements dominate the development of new print media: rapid dry time, customer choices for print gloss, predictable and stable color, and durability in the form of waterfastness and fade resistance.

Inkjet printing is now so highly developed that the consumer can purchase media for almost any application: specialty materials for iron-on transfers and creative projects, greeting cards, business cards, labels, brochures, fine art papers, and papers that enable an inkjet printer to be used for pre-press proofing in commercial printing applications.

4.2 INKJET PRINTING TECHNOLOGIES

4.2.1 THERMAL INKJET

About 75% of inkjet printers sold today employ the thermal inkjet principle. In a thermal inkjet printhead, thousands of drop ejection chambers may be placed on a single silicon chip. Each drop generator consists of a heater, a nozzle, and an ink refill channel, as shown schematically in Figure 4.3. It is common to have one, two, or three colors printed by a single monolithic printhead, and recent developments offer six, eight, and nine colors. With hundreds of millions of thermal inkjet printers in use worldwide, tens of millions of print cartridges and ink cartridges are manufactured every month.

There are two configurations of the thermal inkjet drop generator, called top-shooter and side-shooter, depending on the orientation of the nozzle with respect to the planar substrate. HP, Lexmark, and Canon products employ the top-shooter configuration, whereas the edge-shooter is used primarily by Canon.

In thermal inkjet printheads, nothing moves except the ink itself. This process is shown schematically in Figure 4.4. To eject an ink drop, an electrical pulse is applied for about two microseconds to a thin-film resistor raising its temperature at more than 100,000,000°C per second. Near the heater surface, ink is heated from about 60°C to above 300°C as a metastable liquid, and in a unique thermo-dynamic process it undergoes a superheated vapor explosion. This generates an energetic vapor bubble that grows and collapses to eject a repeatable quantity of ink from the nozzle.

Superheat is a condition where a liquid contains more thermal energy than required for boiling at a given pressure. A superheated liquid can explosively change into a gas when a site is available for a bubble to form (nucleate). This

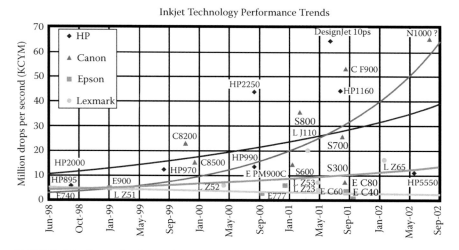

FIGURE 4.3 (See color insert following page 176.) Thermal inkjet configuration.

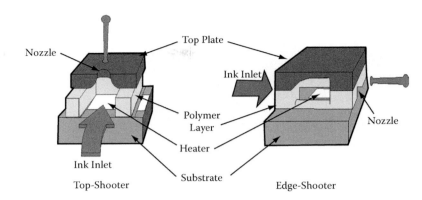

Top Plate

Nozzle

Ink Inlet

Polymer Layer

Nozzle

Heater

Ink Inlet

Substrate

Top-Shooter

Edge-Shooter

FIGURE 4.4 (See color insert following page 176.) Thermal inkjet drop ejection process.

can be a tiny vapor-filled pore in the surface wet by the liquid. For water-based inks, the superheat limit is about 340°C. Above this temperature, the ink cannot exist as a liquid and spontaneously becomes a gas. The actual temperature of vapor bubble nucleation depends on the ink composition, the microstructure of the heater surface, and the size of gas bubbles trapped in microscopic pores on that surface. This process is not boiling in the everyday sense. Boiling in water (and water-based inks) occurs when there are plenty of "large" nucleation sites available to allow the phase change of water from liquid to gas at about 100°C. The high heating rate assures consistent nucleation only from the high temperature sites because the low-temperature sites (activating at 100°C) do not have time to activate. This gives vapor bubbles that are highly consistent in terms of energy release and timing.

The vapor bubble expands rapidly and then collapses in a few microseconds. In this process, it acts like a tiny piston that strokes upward from the surface of the heater to fill the drop generation chamber and then retracts. The expansion phase induces a bulk velocity to the ink, causing a jet to leave the nozzle at 10–15 meters/second. The initial pressure inside the vapor bubble is momentarily greater than 10 atmospheres, but its expansion is so rapid and the quantity of vapor is so small that the positive pressure phase of expansion lasts only about 1 microsecond. During most of the bubble's lifetime of 20 microseconds or less, the pressure inside the bubble is subatmospheric, but the bubble continues to expand to fill about 50% of the chamber carried by the ink it has set into motion.

With no check valve on the drop generator chamber, some ink is forced back through the ink refill channels during bubble expansion. This is useful to dislodge particles and air bubbles in the ink that may accumulate near the inlet structure. Fluidic design of the inlets and a staggered firing order minimize fluidic crosstalk between adjacent nozzles. The fluidic impedances of the nozzle and inlets are balanced to maximize drop ejection efficiency while tuning the drop generator for optimal refill and frequency response.

The vapor inside the bubble has much lower thermal conductivity than the liquid ink, so it insulates the ink from the heater. As a result, little additional vaporization occurs once the bubble covers the heater surface. The electrical pulse is timed to end shortly after vaporization occurs, and this limits the peak temperature of the heater surface and allows it to cool by conduction into the substrate as the bubble expands and collapses. When bubble collapse places ink back into contact with the heater, the surface temperature must be below that required to form a new vapor bubble. This ensures that each electrical pulse produces only a single bubble. There is a thin-film insulating layer, called a thermal barrier, between the heater resistor and the silicon substrate that controls heat flow into the substrate. If it is too conductive, too much heat is lost during the heating phase and this increases the power required to form a vapor bubble and produces excessive printhead heating. If it is not sufficiently conductive, the heater does not cool quickly and multiple vapor explosions can occur. Three or more successive nucleations from a single heating pulse have been observed in this situation.

As the vapor bubble reaches its maximum volume and begins to collapse, the nozzle meniscus retreats into the drop generator, causing the jet of ink to elongate and break off. It takes about 100 microseconds for the drop to travel between the orifice plate and the print medium, where the image is recorded by the ink drops. The carriage holding the printheads typically scans at 0.5–1 meter/second in desktop inkjet printers, so aerodynamic effects can affect drop placement, especially for small drops.

As the bubble collapses, the nozzle meniscus retracts deeply inside the drop generator. The final stage of bubble collapse is so rapid that very high pressures can be created in a cavitation process. Fluidic design of the chamber and selection of materials in the thin films covering the heater ensure that the energy released during final collapse does not damage the heater.

The process of ejecting a drop is also the mechanism to refill the drop generator. Once the bubble collapse is complete, the meniscus is deeply retracted into the drop generator. Surface tension in this extended meniscus creates a subatmospheric pressure in the drop generator to draw ink from the supply reservoir to refill the chamber. As refill progresses, the meniscus returns to a stable position in the nozzle bore near the outer surface of the orifice plate.

Ink in the drop generator is maintained at a subatmospheric pressure equivalent to supporting a column of 2–5 inches of water. This prevents ink from drooling out of the nozzle when it is not printing. It also means that the rest position of the meniscus in the nozzle bore is concave into the oriifce plate, where the curvature of the meniscus and ink surface tension balance the negative head produced by the ink delivery system.

This entire process, from the electrical pulse to stabilization of the nozzle meniscus, takes fewer than 27 microseconds in a drop generator designed to eject 36,000 drops per second.

Important design parameters for the drop generator are the geometry of the ink refill channels, the dimensions of the chamber and thickness of the nozzle, and the shape of the nozzle bore. The important fluidic parameters are ink surface tension and viscosity. During refill, the drop generator acts like a simple fluid mechanical oscillator: the refill channel geometry, ink density, and ink viscosity determine the dominant mass and damping characteristics; the nozzle meniscus is the elastic (spring) element whose stiffness depends on its extension into the drop generator chamber, the nozzle bore, and ink surface tension. If the system is under-damped, the meniscus will oscillate in and out of the plane of the orifice plate before settling. Proper design and tuning produce a critically damped configuration that minimizes the time for the meniscus to stabilize, and this maximizes the operating frequency of the drop generator. Ejecting a drop before the meniscus has settled will affect the ejected volume: the quantity of ink in the drop generator depends on the position of the meniscus, and this provides a variable inertial load to the vapor bubble.

Ink viscosity affects both the dynamics of refill and drop ejection, and it is important to control viscosity for consistent drop ejection. Ink viscosity depends on ink formulation, and it varies significantly with temperature. The temperature of the ink in the drop generators is nominally the same as the printhead substrate. Given the temperatures required to form the vapor bubble, this is somewhat surprising. But, the nucleation process is so rapid that heat from the resistor penetrates only 0.1 micron into a column of ink several tens of microns thick. So, the ink in the drop generator receives heat not from the nucleation event but from contact with the substrate between drop ejection cycles. This has been experimentally confirmed by simultaneous measurement of substrate and droplet temperatures in operating printheads.

Left uncontrolled, the printhead temperature can vary by 35°C or more depending on room temperature, print density, and printing duty cycles. Stabilization of the operating temperature to about 60°C during printing is easily achieved with the same heaters used for drop ejection. Active logic circuits and temperature sensors on the substrate deliver short electrical pulses to the heaters between ejection cycles. These do not produce vapor bubbles but deliver sufficient energy to maintain constant printhead temperature. When high print densities produce excess printhead heating, printing may slow down to manage printhead temperature.

Because a thin film of ink is heated to more than 300°C during drop ejection, thermal inkjet inks must be formulated to avoid *kogation* on the resistor surface.[*] Kogation can be eliminated by assuring purity in the ink ingredients and choosing molecular structures whose thermal decomposition products are soluble in the ink. Although this constraint on ink design may seem at first to be a limitation, thermal inkjet drop generators have a practical lifetime of hundreds of millions of cycles. The technology's ability to use both dye and pigment inks and to produce photographic-quality images and high-quality, durable black text on plain paper speaks for itself.

---

[*] Kogation is derived from the Japanese word *koga*, referring to rice burned onto the surface of a rice-cooking pot, and this term was coined by Canon.

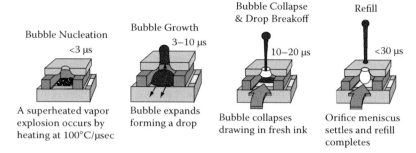

FIGURE 4.5 (See color insert following page 176.) Top view of particle-tolerant ink supply channels.

For high reliability, it is essential to design printheads so that the ink inlets to each ejection chamber are resistant to clogging. Printhead components are carefully cleaned and assembled under cleanroom conditions, but the refill channels are only 10 to 20 microns wide. They can be clogged by particles or fibers that escape cleaning and filtration or are formed during long-term exposure of the print cartridge materials to the ink. Particle-tolerant architectures, shown by the examples in Figure 4.5, prevent inlet clogging by acting as a filter and provide multiple ink refill paths to each drop generator. The dark areas in these photo-micrographs show the photo-imageable polymer thick-film with the orifice plate removed. The light areas are the exposed substrate seen through open channels in the polymer. Note the rows of pillars that are used for particle barriers.

The vertical growth of the vapor bubble in the drop generator is on the order of tens of microns. This is quite large, considering the heater is of similar linear dimensions. This large volumetric displacement right at the nozzle allows the linear packing density of thermal inkjet drop generators to be very high: 600 nozzles per linear inch in a single column (42 microns center-to-center). Printheads usually have two columns of drop generators with a half-row offset between columns. This gives a native vertical printing resolution of 1200 dpi. Drop ejection timing allows such printheads to place drops at 4800-dpi grid points along the horizontal (scan) axis, and very smooth lines and character edges can be produced with a 4800- by 1200-dpi print grid. This level of fine addressability is very useful for color halftoning in image reproduction.

A silicon substrate allows thermal inkjet printheads to be fabricated using techniques employed in the manufacture of integrated circuits. The details of typical thin- and thick-film layers are shown schematically and not to scale in Figure 4.6. For reference, the thin-films are about one micron deep and the thick films and orifice plate form a stack about 50 to 70 tall. The heaters used in drop ejection are thin-film electrical resistors sputtered over a thermal barrier to control heat conduction into the substrate. The heaters are covered with dielectric and metallic layers to resist chemical attack and cavitation. These films are designed for high fracture toughness, which is achieved by sputtering materials with very

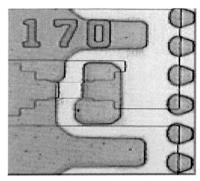

10 Picoliter   4 Picoliter

FIGURE 4.6 Thin- and thick-film structure of a thermal inkjet printhead.

low brittleness. Low-resistance metallic conductors and active electronic elements drive the heaters and demultiplex signals from the electrical interconnect.

After fabricating the thin-film layers, ink channels and the drop ejection chamber are formed, applying a photo-imageable polymer thick film tens of microns thick to the substrate and thin films. This thick film is exposed to light through a pattern mask and then chemically developed to remove material, forming chambers, channels, and pillars. The orifice plate, containing the inkjet nozzles, is then attached to complete the structure.

Two technologies are commonly used for orifice plates: electroformed thin metal plates and laser-ablated polymer sheets. Nozzles must be very precisely located with respect to the heater and chamber walls; consistent in length, diameter, and bore shape; and free of defects that could perturb or deflect the droplets. Nozzles producing 4-picoliter drops are typically less than 15 microns in diameter: about 1/3 the diameter of a human hair.

Electroforming is an electrochemical plating technique that deposits a thin metal layer (called a foil), usually nickel, onto a substrate with a patterned metal layer on top of an insulator. The metal pattern defines the plating areas, and circular dots exposing the underlying insulator define the nozzles. Other geometries of exposed insulator define openings in the foil useful for alignment targets and separating the individual orifice plates. Hundreds of orifice plates are formed simultaneously on the foil sheet, each with hundreds of nozzles.

As the plating process adds depth to the foil, metal advances uniformly across the insulator dots from the edges. This creates a smooth converging nozzle bore. The final nozzle diameter depends critically on the plating thickness because the diameter grows smaller at twice the rate of depth accumulation.

The material composition of the foil and the patterned metal are chosen such that the foil has weak adhesion to the substrate. This property allows it to be peeled off after plating is completed. A thin gold passivation layer is often electroplated onto both sides of the foil to minimize corrosion from the ink, and the individual orifice plates are then separated from the sheet and assembled onto the substrate's thick-film layer.

Laser ablation uses an excimer laser delivering bursts of pulses, each of a nanosecond duration. Each pulse vaporizes a thin, patterned region in a polyimide (Kapton®) tape to open a hole or channel. Multiple pulses are required to ablate through 50 microns of polyimide. The vapor produced by the ablation process carries away the ash to leave a clean hole. The effects of laser ablation are highly localized, and there is virtually no thermal damage to the material surrounding the nozzle. Compared to the multiple process steps and wet chemistry of electroforming, laser ablation is a dry process offering an economical, high-productivity method to produce a sharp-edged nozzle bore. Another key advantage of this method is that the polyimide tape can function both as the orifice plate and the electrical interconnect.

Figure 4.7 shows a 512-nozzle thermal inkjet printhead with a silicon substrate and laser-ablated orifice plate integral to a polyimide tape. This tape has electroplated electrical conductors connecting the substrate to the 21 gold-plated electrical interconnect pads. These pads provide power, control signals, and data interconnection between the printhead and printer. Control, demultiplexing, and drive electronics integrated into the printhead substantially reduce the number of electrical interconnect pads, and this minimizes the cost and size of the printhead and improves the reliability of electrical interconnect.

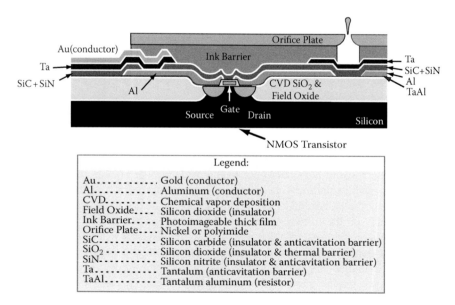

FIGURE 4.7 (See color insert following page 176.) An HP thermal inkjet printhead and electrical interconnect.

Considering the leverage from integrated circuit manufacturing and process control, thermal inkjet offers a very cost-effective method for high-volume production of inkjet printheads.

4.2.2 PIEZOELECTRIC INKJET

About 25% of inkjet printers sold worldwide use piezoelectric transducers for drop ejection. Piezoelectric materials change their dimensions in response to applied electric fields, and the most common material used in piezo inkjet printheads is lead zirconate titanate (PZT) ceramic.

When driven by an electric pulse, the piezo element typically changes its long dimension or bends. In either case, the displacement is typically a micron or less. Whereas thermal inkjet can be characterized as a piston (vapor bubble) stroking in a cylinder (drop generator chamber), the piezo drop generator operates on the principle of a volumetric transformer: a small displacement (piezo deformation) over a large area (diaphragm) produces a large displacement (the ink droplet) over a small area (nozzle). To work properly, the fluid in the drop generator must be practically incompressible.

Epson's MLP drop generator, circa 1997,[1] shown in Figure 4.8 is an example of a device operating by direct elongation of a piezo crystal. In this printhead, each ejection chamber is formed in a silicon channel layer with a stainless steel orifice plate on one side and a diaphragm on the other. A 5-mm-long PZT transducer bonded to its exterior displaces the diaphragm wall of each chamber.

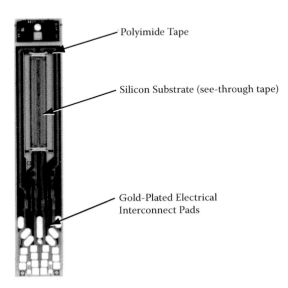

Polyimide Tape

Silicon Substrate (see-through tape)

Gold-Plated Electrical
Interconnect Pads

FIGURE 4.8 (See color insert following page 176.) Epson MLP piezo inkjet drop generator.

The diaphragm seals the chamber, allows displacement of the wall, and isolates the PZT ceramic from the ink. The PZT transducer is fabricated from a stack of 20-micron-thick layers to reduce the required drive voltage. Dimensional changes are small: in operation, the 5-mm-long transducer elongates by only one micron. As a result, the drop generator chamber requires a large diaphragm to create the required volumetric displacement, and this makes the piezo drop generator much bulkier than a thermal inkjet producing a comparable drop volume. Nozzles are spaced 140 microns apart, giving a linear density of 180 orifices per inch. This design prints dot rows at 180-dpi vertical resolution, so the complete area fills at 720 dpi require 6 passes where the paper is advanced 1/720 inch between each printhead scan.

Rather than using direct elongation, many piezo printheads employ a bending-mode bimorph sandwich using a thin piezo element bonded along its length to a thin piece of another material. The Epson MLChips printhead,[2] shown schematically in Figure 4.9 is an example of such a design with a PZT/zirconia bimorph attached to a zirconia ink chamber. Three laminated stainless steel plates forming the nozzle layer, ink manifold layer, and ink inlet layer are adhesively attached to the zirconia layers to complete the structure. The bimorph structure bends when voltage is applied to the piezo element, and this produces a reduction in the chamber volume to eject a droplet of ink. Bimorph deflection in this design is about 0.1 microns, so the lateral dimensions need to be larger than in the MLP design to achieve the desired volumetric displacement. Chamber width is 340 micron, compared to 100 microns for the MLP design, and the nozzle spacing is therefore larger. This printhead can be produced at lower cost than the MLP, and high printing resolution is achieved with multiple passes.

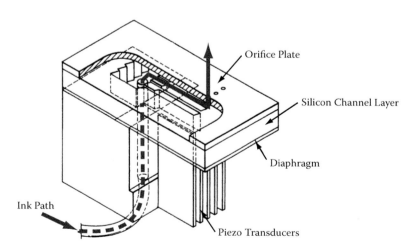

FIGURE 4.9 Epson ML chips drop generator.

A bending-mode bimorph configuration is also used in the solid inkjet Xerox Phaser printheads.[3] These printheads are fabricated entirely from a stack of photomachined stainless steel plates, brazed together with thin gold bonding layers. Photochemical machining typically uses a dry-film resist and spray etchants to produce plates with holes and channels.

A variety of electrical waveforms is used to drive piezo inkjet printheads. This is an advantage of piezo inkjets that allows a given design to deliver the same drop volume using inks of different viscosities and surface tensions, to adjust for manufacturing variations, and to eject drops with different volumes from a single orifice.

A common waveform is the bipolar drive pulse, in which a pulse of reverse polarity is first applied to the piezo transducer to enlarge the ejection chamber. This condition is maintained until ink has been drawn into the chamber from the ink reservoir. Then a normal polarity pulse is applied to reduce the chamber volume and eject the ink from the nozzle. Compared to a unipolar pulse, a bipolar pulse with the same peak voltage can eject higher drop volumes.

More complex waveforms enable piezo inkjets to eject drops of different volumes from an orifice, as seen schematically in Figure 4.10. This technique uses a first pulse to excite an oscillation of the meniscus in the nozzle, followed by a higher-amplitude pulse (or pulses) to eject the drop. Depending on the relative timing and amplitude of these pulses, 2.5-, 5-, and 11-pl drops can be produced.

One inherent disadvantage of piezo inkjet is its high sensitivity to air bubbles. Unlike thermal inkjet, in which the pistonlike vapor bubble produces high ink velocities throughout the drop generator to sweep out air bubbles, the ink in a piezo drop generator is virtually at rest except where the chamber cross-sectional area is very small compared to the diaphragm: near the orifice. This makes it difficult for piezo inkjets to purge bubbles in normal operation. Bubbles trapped

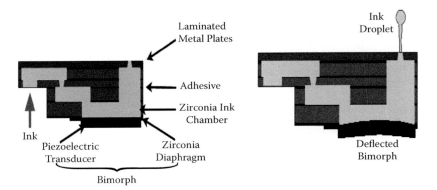

FIGURE 4.10 (See color insert following page 176.) Drive waveforms for variable drop volumes.

in the drop generator can completely disable the piezo drop ejection process because it critically depends on the incompressibility of the ink in the drop ejection chamber. Bubbles are compliant and reduce the amplitude of the volumetric displacement at the nozzle.*

To prevent performance degradation from bubbles, piezoelectric printheads require regular priming by pumping ink through the printhead and disposing of it in the printhead service station. The large waste ink disposal pads found upon disassembly of piezo inkjet printers indicate the large quantity of ink required for this process over the lifetime of the printer..

A key difference between piezo and thermal inkjet is that the piezo drop ejection process does not heat the ink. This allows piezo inkjets to use ink components and colorants that would be unsuitable for use in thermal inkjet devices. So far, this has not proven to offer any practical advantage, as inks have been developed for both technologies that achieve similar imaging performance in desktop color printing applications. Ink compatibility with the materials used in the printhead and ink delivery system is by far the major determinant of the range of solvents and other ink components that can be used by a particular technology.

Some disadvantages of piezo technology over thermal inkjet include higher fabrication cost, lower nozzle density, larger overall size, and greater sensitivity to trapped bubbles.

Piezo printheads are designed to last the lifetime of the printer and are typically not user-replaceable. Kogation and cavitation are not failure mechanisms for piezo printheads. Piezo and thermal inkjet printheads are both subject to degradation from long-term exposure to ink, and this can cause failure of adhesives and structural delamination, changes in nozzle shape, unrecoverable clogs, corrosion, and material property changes.

4.3 INK STORAGE AND DELIVERY

The early history of unreliable products with messy liquid ink supplies created significant barriers to adoption for first generations of modern inkjet printers. Providing clean, simple, and economical replacement of consumables would seem to be the whole story for the ink delivery system. To understand the success of desktop color printing is to recognize the impact the ink storage and delivery system has on consistent print quality and reliable operation and how it is one of the most important determinants of printer manufacturing cost and the cost to print a page.

---

* This effect is seen in a symbolic volume conservation (continuity) equation for the drop generator: (diaphragm stroke × diaphragm area) = (total change in trapped bubble volumes) + (ink volume backflow into the refill channels) + (ink displacement × nozzle area). A non-zero first term on the right-hand side of the equation obviously reduces the subsequent two terms. This is only a kinematic equation that does not take into account viscous and inertial effects in the nozzle and refill channel, and these set the ratio between the second and third terms.

Desktop color inkjet printers have multiple printheads mounted side by side on a carriage that scans across the paper. The printheads print rows of dots, the paper is advanced, and the process repeats. The print carriage must be able to move out of the print zone on either side of the scan to allow all printheads to print up to the left- and right-hand margins; to allow the carriage to decelerate, stop, accelerate back across the print zone; and to park the carriage in the printhead service station. Compact printers fulfill an important user need for the efficient use of desktop space. The width of the printer is driven by the widest media it can handle plus about twice the width of the print carriage. Minimizing the width of the carriage is an important design consideration, because printer width has a strong influence on manufacturing cost. For example, manufacturing costs increase for a wider printer that requires a longer print carriage slider rod, longer electrical cables to the carriage, a longer encoder strip, a longer drive belts, and the need for stiffer plastic and sheet metal parts in the chassis.

High printing throughout depends on minimizing the time the printheads are not printing. This means that the print carriage must stop quickly at the end of a scan and rapidly accelerate to constant velocity (typically 0.5–1 meter/sec) back across the print zone. To minimize printer width, an acceleration of 0.5–1 g* requires the printer to deliver a force on the print carriage equal to 50–100% of its weight. It is easily appreciated that minimizing carriage mass reduces the cost of the scan axis servo motor and power supply, reduces an annoying side-to-side shaking as the printer operates, and may even make the printer quieter.

The amount of ink onboard the carriage and principle of its containment and pressure regulation directly affect print carriage width and mass. The amount of ink contained in a disposable print cartridge has a direct impact on the cost per page. But, the actual ink yield is a function not only of the initial fill quantity but also how much ink is used to print, how much is consumed during printhead service cycles, and how much is unusable due to ink delivery system characteristics.

In a drop-on-demand printer, hydrostatic pressure in the drop ejection chamber is maintained at a slightly negative pressure relative to the atmosphere to produce enough suction to support a water column 2-5 inches high. This prevents ink from flowing out of the nozzles when the printhead is idle. Air is not sucked into the printhead because the nozzle menisci act like elastic membranes whose deformation (into the nozzle bore) balances this negative pressure. Shock to the printhead, such as dropping it on a hard surface, can cause the menisci to detach from the edge of the nozzle and allow air to enter causing a deprime.

Consistent drop volumes depend on maintaining constant delivery pressure at all flow rates over the life of the ink supply. Typical desktop printer ink supplies hold between 7 and 69 ml. In some designs, the ink delivery rate to a single printhead can vary between zero and a peak value of 7 ml/minute.

A number of different technologies are commonly used to store and deliver ink to the printhead. Ink reservoirs moving with the print carriage, either integral with a printhead or separately replaceable, are called on-axis supplies, meaning

---

* 1 g is the acceleration of gravity: 9.8 meters/second/second.

"on the scan axis of the print carriage." Fixed ink reservoirs separated from the printheads are called off-axis supplies. The following discussion begins with various types of on-axis ink supplies.

Early inkjet products used elastomeric bladders. Starting from a dimple, bladders are designed to collapse in a manner that applies a nearly constant (suction) pressure as ink is withdrawn. The need to ensure a repeatable collapse limits bladder capacity to a few milliliters of ink. The bladder must provide a barrier to gases and volatile ink compounds. Material compatibility with the ink is a significant design issue for bladders: vulcanizing agents and mold release compounds can contaminate the ink and dramatically affect physical properties such as surface tension. Bladders are usually hemispherical or cylindrical, and these bulky configurations limit the usefulness of bladders when a higher-volume, compact ink delivery system is required.

A large percentage of disposable thermal inkjet print cartridges hold the ink in a block of non-rigid, open-cell foam inside a rigid plastic container. Foam provides a porous medium for ink storage and flow. The capillary action at the ink–foam–air interface maintains suction as ink is withdrawn. Foam ink cartridges have a small vent to allow air to replace the ink as it is consumed. The pore size and compression of the foam must be carefully controlled to produce a nearly constant delivery pressure over the life of the ink supply.

Given the large surface area it presents to the ink, ink–foam material compatibility presents a major issue whenever new inks are developed: the foam must be free of compounds such as plasticizers and surfactants, that could interact with ink components. Otherwise, significant changes in ink surface tension can occur, leading to loss of pressure regulation and allowing ink to drool out of the nozzles.

Rapid withdrawal of ink from a foam reservoir can cause a large increase in suction. This can deprime the printhead in severe cases and modulate the drop volume in others. In addition, air can enter the pores and isolate ink-filled regions. Unless these regions reconnect by capillary action, significant quantities of ink can be stranded and the full capacity of the ink cartridge cannot be utilized.

In foam blocks with linear dimensions of more than a few inches, hydrostatic forces can overcome capillary forces. In this case, ink can drool out of the orifice plate in the normal printing orientation or can migrate away from (and possibly deprime) the ink supply pipe to the printhead when the ink cartridge is handled. Long ink paths to the supply pipe increase the possibility of stranding ink. For these reasons, and to minimize the linear dimension of the ink cartridge, foam systems are useful up to about 20 ml of ink per color.

Some simple mechanical systems, including bubblers, spring bags, ink bags, and regulators, have been developed to overcome the limitations of bladders and foam.

A bubbler uses a small orifice and a labyrinth baffle to allow air to bubble into a rigid chamber holding liquid ink. This produces a controlled suction in the ink supply as ink is consumed. A small spring-loaded plastic bag in the chamber, vented to the atmosphere, acts like a bellows to balance changes in atmospheric pressure. Bubblers can supply about 40 ml of ink, twice as much in the same (nearly cubical) volume as a 20-ml foam-based ink cartridge.

Spring bags contain the ink in a sealed, flexible, metalized plastic bag. The bag contains two metal plates held apart by a leaf spring. This mechanism is immersed in the ink, and the plates press outward against the walls of the bag to resist its collapse as ink is withdrawn. This allows a large quantity of ink, up to about 100 ml, to be withdrawn at nearly constant pressure. The bag is contained within a rigid cartridge body. Spring-bags allow the design of a narrow and tall ink cartridge, and the collapse of the bag can operate a simple mechanical ink quantity indicator on the print cartridge. Spring-bag ink cartridges have been used in more than 100 million HP DeskJet and PhotoSmart printers sold since 1993.

Metalized plastic bags of ink contained in rigid plastic cartridge bodies are used in off-axis ink delivery systems. Only the space available and industrial design of the printer limit the capacity of this system, which delivers 69 ml of ink per color in some configurations. Ink is pumped through pressurized flexible tubes from the ink bags to the printheads. The tubes are designed to limit evaporation of volatile ink components, provide a barrier to air, and remain flexible for the life of the printer. A pressure regulator on or near the printhead admits pressurized ink as it is consumed and delivers it to the printhead at a constant negative pressure. Some ink cartridges also contain a pad to absorb waste ink produced during printhead servicing.

4.4 PRINTHEAD SERVICE AND MAINTENANCE

Desktop inkjet printers require service stations to ensure reliable operation of the printheads. Service stations perform the essential functions of capping, spitting, priming/purging, and wiping. These functions are controlled by printer firmware that keeps track of drops printed by each nozzle on each printhead, the time the printhead is exposed to air while printing, the longest time any nozzle has gone without printing, the number of pages printed between service cycles, the time since the last page printed, and other operational parameters.

Some sophisticated service stations use optical or electrostatic drop detectors to determine which nozzles are operating within drop weight, drop velocity, and trajectory specifications. This information is used in nozzle substitution algorithms to replace unfit nozzles by good ones in multiple-pass print modes until the nozzles are either recovered by a service cycle or identified by the printer as unusable.

An elastomeric cap in the service station contacts the orifice plate of each printhead to seal the printhead nozzles from the atmosphere when not in use. The cap provides a humid environment that limits the evaporation of volatile ink components, preventing the formation of viscous plugs and ink crusts.

Viscous plugs and ink crusts can make drop ejection impossible, and this problem is more severe with piezo inkjets than thermal inkjets due to piezo's less energetic drop ejection process. Ink crust at a nozzle's edge can misdirect ejected drops. With some inks, the ink crust, once formed, is insoluble in the ink vehicle.

Allowed to harden, it may be impossible to remove and the printhead must be replaced. A soft crust may be mechanically removed by wiping. Capping provides a critical function: for some inks, the nozzles can be exposed to air for only a few seconds without forming crusts or viscous plugs.

Along with viscosity increases, the loss of volatile ink components from the nozzles can concentrate colorants. If the nozzle is actually able to eject this concentrated ink, which will have different fluid properties, then its printing characteristics will be different from normal ink. The dots may be misplaced and have higher optical density. Different spread and penetration characteristics on the print medium will produce a different-size dot. For this reason, it is important to keep fresh ink in the nozzles, especially when certain nozzles or colors have not printed for a while. For example, when printing a black text document, the color printheads are exposed to air in the scanning carriage but may not print for several minutes. However, they must be ready to produce drops meeting print quality specifications at any time. This is assured by a process known as "spitting."

Spitting periodically ejects drops from each nozzle into the service station's waste ink reservoir. Spitting a few drops from each nozzle can eliminate concentrated colorant and viscous plugs developing in unused nozzles. If a printer is observed to pause in the middle of a page to return the print carriage to the service station for a few seconds, it is almost certainly to spit from unused nozzles.

Priming and purging are processes that remove air bubbles trapped in the drop generators. A pump is required to create a subatmospheric pressure (or impulse) to draw ink out of the printhead. Suction is preferred to pressurizing the drop generators through their ink supply channels, because suction causes trapped air bubbles to expand and be drawn out by ink flow. The amount of ink used in servicing the printhead can be substantial, especially for piezo printheads, which do not have an effective means for continuously purging the drop generator of trapped gas bubbles.

The ink used in spitting, priming, and purging must be stored in a waste receptacle designed to last the lifetime of the printer or in a waste ink container in a disposable ink cartridge. This is typically an absorbent pad connected by tube or capillary to the service station spittoons and priming pumps. Evaporation of volatile ink components over time, especially water, helps limit the volume of the waste ink storage medium.

A flexible, elastomeric wiper blade can be passed across the orifice plate to remove ink spray, ink crust, and paper dust. In some cases, the printheads eject a small amount of ink on the wiper to obtain a wet-wipe, and this helps to dislodge ink crusts.

Materials compatibility between the inks and service station caps and wipers must be assured. Otherwise, the elastomer may decompose, leaving sticky residues on the orifice plate that permanently clog nozzles. In the worst case, these residues can affect the wetability of the orifice plate, causing a complete loss of ink containment with the potential of irreparable damage to the printer and its environment. Apart from image quality, the consequences of material incompatibilities are the reason why most inkjet manufacturers recommend their own original ink supplies and caution against ink cartridges supplied or refilled by third parties.

Wiping multicolor printheads, where two or three different inks share the same orifice plate, can pose problems from ink mixing. Liquid ink from one color, or from spitting on the wiper, can be drawn by capillary action and negative head into drop generators of another color. This can produce off-color droplets or chemical reactions that clog the nozzles. In this case, spitting occurs during or immediately after the wipe to ensure ink purity in each color printer.

Excessive wiping can scratch the orifice plate and produce capillary channels drawing out ink and accelerating crusting. Wiping can damage non-wetting coatings and wrinkle a flexible plastic laser-ablated orifice plate causing it to delaminate from the underlying structure. In the worst case, wiping can exacerbate nozzle clogging by forcing foreign materials into them. For these reasons, wiping is done only when necessary. Some service stations keep track of the number of wipe cycles for each printhead in order to balance the benefits and harms of wiping.

The user can initiate cleaning and priming service cycles if the print quality becomes unsatisfactory (e.g., white streaks are seen in characters or area fills), using commands in the printer driver or toolkit installed with the driver on the user's computer.

4.5 INKJET INKS

Both liquid and solid (hot-melt) inks find application in desktop color inkjet printers, with liquid ink used in virtually all consumer and business inkjets and solid ink in some graphic arts and proofing applications.

Whether liquid or solid, ink has two functional components: the colorant and the vehicle. Although the role of the ink vehicle would appear simply to deliver colorant to the paper, in fact, ink touches everything in the printing process and it is the most functionally complex part of any inkjet device.

4.5.1 LIQUID INKS[*]

The journey of a liquid ink drop begins with the storage of ink over a shelf life of 6 to 54 months and a service life of many months in the printer. During this time, the colorant must remain dissolved (dyes) or suspended (pigments). Chemical and physical properties must be stable without chemical interactions between ink components and materials in the ink containment, pressure regulation, and printhead components. Ink surface tension may play a role in regulating pressure in the ink delivery system. Service station features and functions depend on ink volatility and crusting characteristics. While the nozzles are exposed to air, ink

[*] The authors acknowledge contributions to this section by Dr. Nils Miller and Dr. John Stoffel of the Hewlett Packard Company.

surface tension and viscosity must remain within narrow limits to ensure proper drop volume, velocity, and trajectory. In a thermal inkjet, the ink's vaporization characteristics are part of the drop ejection process, and any thermal decomposition products must remain soluble in the vehicle. Once the droplet reaches the paper, the vehicle must rapidly wet and penetrate the surface to control dot spread. While penetrating, the vehicle separates from the colorant, leaving it near the surface to maximize optical density. The vehicle should minimize cockling of plain papers and evaporate, leaving behind a dot that meets print-quality objectives for size, shape, optical density, fade resistance, and waterfastness.

Typically, ink vehicle in desktop color printers is mostly water with various additives. The colorant is only about 5 to 10% of the ink by weight. Although water is not necessarily the ideal solvent from printing considerations, especially on plain papers, it is the only solvent that meets worldwide health and safety regulations for the home and office. These regulations also set guidelines for flammability and toxicity and limit the percentages of organic compounds, such as alcohols and glycols, that may be used to meet various performance requirements.

The formulation of inks represents significant intellectual property developed over years of research and development, covering thousands of compounds and combinations. Introducing a new ink can cost millions of dollars exclusive of investment in production facilities. Although manufacturers may have hundreds of patents issued on ink, certain properties, ingredients, and manufacturing processes remain trade secrets. Some essential components in the finished ink are difficult to characterize, even with the most modern analytical chemistry techniques. For this reason, inkjet inks are difficult to completely reverse-engineer.

Typically, inkjet manufacturers partner with chemical companies to research and develop new colorants and other key ink components, and the inks are formulated by the printer manufacturer to be part of a system with the printer, printheads, and specialty media (such as photo papers). Once the ink properties are fixed for a specific product platform, the partner builds and operates the ink production facilities. To give a sense of scale, ink production began at 250,000 liters a month for a new print cartridge platform recently introduced by a major inkjet manufacturer. The ink was delivered to the print cartridge production line in tanks each containing a cubic meter (1000 liters).

Before printheads or ink cartridges are filled with ink, the ink undergoes hundreds of physical and chemical analyses in the manufacturing process to ensure consistent properties and performance. These analyses include measurements of pH, viscosity, surface tension, and particle size distribution; gas and liquid chromographic analyses of solvents and dyes; spectrographic elemental analysis; anion analysis; testing in printheads for drop ejection consistency, storage and operational life, and materials compatibility; and print testing for smearfastness, fade, waterfastness, color, and optical density.

The largest component of ink is typically highly purified water, and most inks used in desktop inkjet printers have a viscosity of between 1.5 and 3 centipoise. Surfactants are added to obtain a surface tension in the range of 20 to 30 dynes/cm. Humectants, usually glycols, may be added to minimize the effects of viscous plugs in the nozzles by maintaining equilibrium between water vapor loss to and absorption from the atmosphere. Small amounts of alcohol increase the rate of vehicle penetration into the paper by increasing the ink's ability to wet the surface. Cosolvents keep dyes in solution and pigments in suspension. Buffering agents maintain ink pH within a narrow value. Biocides and fungicides prevent growth of organisms in the ink during storage and use. Agents are added to control cockle on plain papers.

The payload of the ink vehicle is the colorant. All manufacturers of desktop color inkjet printers offer both dye and pigment colorants. Each type has its advantages and disadvantages and is ideal for certain applications and somewhat less suited to others.

4.5.2 DYES

Dyes are molecules designed to absorb specific frequencies of light, and they are chemically dissolved in the ink vehicle. With dimensions on the order of a nanometer, dye molecules are small enough to penetrate into any absorbent medium, so the surface of the print medium determines the gloss. Because they are too small to scatter light, dyes are typically brighter and more colorful than pigments, and this usually allows the dye-based printer to produce the largest color gamut.

Dyes are individual molecules, and after printing their internal chemical bonds can break due to exposure to light, moisture, oxygen, and other environmental chemicals. The interaction of dye molecules with the print medium has a dramatic effect on the fade resistance and waterfastness of the print. Encapsulation of dyes within a coating on the print medium mitigates all effects but light exposure, and UV-absorbing laminates can significantly extend the time to fade. For example, a certain set of inks printed on photo papers specifically designed to resist fade can deliver more than an order-of-magnitude longer fade resistance than on other media. Similarly, some dye-based inks may deliver poor waterfastness on some media, but others can provide a completely waterfast solution.[*]

One method for improving fade resistance is to design the primary colorants to fade at the same rate. When one primary, usually magenta, fades faster than the others, an objectionable hue shift will occur. This is commonly seen in conventional photographs after long-term exposure to florescent lights or sunlight: skin tones become greenish with the fading of the photographic magenta dye. When all the dyes fade uniformly, the decrease in color saturation is far less noticeable to the eye than a hue shift.

4.5.3 PIGMENTS

Pigments are particles about 50 to 150 nanometers in diameter composed of tens of thousands of dye molecules bound together. Unlike dyes, pigments are not dissolved in the ink vehicle: they are formulated with dispersing agents to create

---

[*] See the discussion of porous and encapsulating coatings in the section on inkjet media.

a stable suspension. If this suspension fails during storage or from exposure to temperature extremes, contaminants, or printhead materials, the pigments will come out of suspension in the ink vehicle to form sludge in the ink container or printhead. This is called a pigment "crash." This is a failure that can be recovered only by replacing the ink supply, printhead, or both. It is why it is important to observe the "Use By" date the manufacturer specifies for the ink cartridge. Pigment ink cartridges from some manufacturers have a shelf life of only six months.

Pigment particles are too large to penetrate most inkjet papers or photo media. Instead, they form a thin film on the surface, and it is difficult to achieve a high, uniform gloss meeting photo-quality expectations: the printed regions can appear dull when viewed at certain angles. Some manufacturers offer special media for pigment inks that provide a uniform satin gloss.

Although pigment inks for desktop printers are water-based, pigments offer excellent waterfastness and resistance to highlighter smear on plain papers by forming stable chemical bonds with cellulose fibers. Waterfastness usually takes several hours to fully develop. Pigments can offer excellent fade resistance: when dye molecules on the surface of the particle decompose, underlying molecules take their place. The more chromatic pigments tend to be less fade resistant.

The most recent generations of high chroma color pigments for photo applications offer fade resistance in the 100-year range according to predictions from industry-standard test methods, such as those performed by Wilhelm Imaging Research.

In some cases, dyes are added to pigment inks to achieve higher chroma or to adjust the hue. Fading of these dyes over time or reduced waterfastness can undo their benefits.

The chemistry of pigment dispersing agents can be designed to react with dye-based color inks so that mixing the two on the print medium will cause the pigment to crash. This achieves high edge sharpness in black text or lines printed on a colored background, particularly a light color such as yellow, because the feathering of black text into light colors produces an objectionable loss of sharpness.. When printing on plain paper, a process known as "black fortification" prints a small amount of dye-based cyan and magenta inks under the areas to be printed with black pigments. The pigment particles are quickly immobilized before capillary forces can carry them away, feathering the dot edge. This allows some inkjet printers to produce uniformly high quality black text on a variety of plain papers, and this media-independent print quality meets a significant user need.

This type of reactive pigment/dye system is incompatible with printing continuous-tone color images. Cyan, magenta, yellow, and black (CMYK) printers using pigment black ink print color images on photo media with only the cyan, magenta, and yellow (CMY) dyes. Overprinting cyan, magenta, and yellow drops at the same location produces black dots. Even disregarding chemical reactivity, the different surface penetration characteristics of dyes and pigments would produce unacceptable non-uniformity in gloss.

Some pigment ink printers use two black inks: one for satin or glossy photo media and one for matte surfaces, such as fine art papers. These inks match the gloss of the substrate, and the matte black usually delivers a higher optical density.

With a diameter between $1/10$ and π the wavelength of green light, pigment particles are large enough to scatter visible light, so they produce less saturated colors than dyes. However, even the smallest inkjet nozzles are 100 times larger than a pigment particle, so pigmented inks still behave like liquids at this scale.

4.5.4 SOLID INKS[*]

Solid inks are also called hot-melt inks and phase-change inks. They are based on materials that are solid at room temperature but have low-enough viscosity in the liquid phase to be jettable. The ink vehicle is typically a natural or synthetic wax or a mixture of both. These materials have glass transition points between 80 and 100°C and are heated to around 130°C to bring their viscosities below about 20 centipoise for ejection.

Once printed, cooling of the ink droplets produces a rapid increase in viscosity followed by a phase-change back to solid form. The ink forms a thin layer on the surface of papers and other substrates, and it does not penetrate like liquid ink. This minimizes the effect capillaries can have on dot size and shape, and keeping the colorant on the surface produces high print density and color saturation. For these reasons, solid inkjets have the potential to deliver print quality with higher media independence compared to liquid ink systems. Despite these advantages, the mechanical properties of solid inks at room temperature present a number of practical challenges. If the layer is too brittle, bending or creasing the paper will cause it to crack. If the ink is made ductile to minimize cracking, it may be too soft and susceptible to scratching. Scratches dull the surface gloss of printed regions and can occur when printed sheets rub against each other in a stack.

Mechanical durability of printed text and images improves with higher glass transition temperature, T_g. A higher T_g protects the printed dots from smearing if they are exposed to high environmental temperatures (such as in a closed vehicle parked in the sun). However, the temperature at which the melt has an ejectable viscosity is typically 20 to 30°C above T_g, and a high T_g presents a problem for using piezo materials. Piezo devices must be operated below their Curie Point,[**] and this presents a constraint of about 130°C on the melt-temperature for solid ink printers.

Solid ink is usually supplied in the form of a solid rod or block, and this is inserted into the printer and melted in a chamber close to the printhead. Xerox keys the cross-sectional shape of each color ink block to that color's re-supply slot thus ensuring that inks are always loaded properly. Some solid ink printers

---

[*] The authors acknowledge contributions to this section by Dr. Wayne Jaeger of Xerox Corporation.
[**] At temperatures above the Curie Point, piezoelectric transducers lose their ability to produce dimensional changes with applied electric fields.

allow ink blocks to be reloaded on-the-fly, and this is a useful feature that allows print jobs to continue without interruption.

Melt-on-demand systems are preferred over systems that keep the ink in liquid form, because keeping the ink hot for long periods can produce changes in material properties due to polymerization, decomposition, and loss of volatile compounds. On the other hand, repeated freeze–thaw cycles, in which the printer is completely turned off between uses, can produce significant mechanical stresses on the printheads due to specific volume differences between the solid and liquid phase. Freezing and thawing can also affect the solubility of the components of solid ink.

Because solid ink printers use piezo drop generators, removing dissolved gases and bubbles from the liquid phase presents a significant challenge. Manufacturers such as Xerox have developed ingenious schemes to extract gases from the melt near the printhead using, for example, a gas-permeable membrane between the melt and a vacuum chamber.

The phase-change process is so rapid that drops form nearly hemispherical beads on the surface. Without a post-treatment to flatten the solidified drops, these beads could give an undesirable grainy texture to text and area fills. Acting as tiny lenses, the beads would scatter light to make solid ink useless for printing color on overhead transparency films. Xerox Phaser products print solid ink drops onto a heated intermediate transfer drum. The image on the drum is transferred under pressure to the print medium, and this produces a smooth surface. Other solid ink printers use a cold pressure roller or re-melt the ink after printing to form a smooth film.

4.6 INKJET MEDIA*

The media available for desktop color inkjet printers present the user with broad choices for meeting cost and quality objectives in document and photo formats. Special media are available for printing banners, brochures, greeting and business cards, iron-on transfers, and other creative projects.

Each major inkjet printer manufacturer develops its own ink formulations, and because inkjet print quality depends on the print medium, inks and ink-receptive coatings are generally designed together for best performance. Therefore, major manufacturers offer a collection of special media optimized for use in its own printers. Because all manufacturers endeavor to meet similar performance and quality objectives, there are enough similarities in their solutions to support third-party suppliers of inkjet papers and specialty media in markets where low-cost and unique features are more important than ultimate image quality or permanence.

The print medium has many requirements for mechanical and imaging performance. Mechanical performance relates to handling and reliability in the

---

* The authors acknowledge contributions to this section by Dr. Nils Miller of the Hewlett Packard Company.

printer paper path and involves material properties of the surface and substrate. Imaging performance relates to optical properties of the surface and substrate and to physical properties that control the spread, penetration, and absorption of ink.

4.6.1 Mechanical Properties

Surface finish and coatings, which control the coefficient of friction with the printer's feed and drive rollers, and the medium's bending stiffness are the most mechanical properties of print media. These determine the reliability of picking single sheets from the input tray and feeding them without jamming or skew through the printer paper path.

Cockle is an uneven, bumpy texture across areas of high print density. Plain papers can cockle in wet regions, because water causes cellulose fibers to swell and water breaks the hydrogen bonds between fibers. This allows fibers to move relative to one another, relieving mechanical stresses built-in during paper manufacture. Cockle can form immediately in the print zones (wet cockle) or in the output tray (stacker cockle), where moisture lost by neighbor sheets causes unprinted regions to cockle. Cockle can partially or completely disappear as the sheet dries, but in the worst case leaves permanent wrinkles that can render the print unusable. Water-based inks usually contain anti-cockle agents to minimize this effect. Severe wet cockle produces wrinkles out of the print plane that can cause printheads to crash against the paper. The printhead-paper distance is typically 1–2 mm, and wet cockle can reach this height. A printhead crash smears ink across the sheet and can damage the printhead's orifice plate. Cockled sheets can cause feed problems during duplex printing. Curl is another moisture-related phenomenon, in which differential expansion (or contraction) of paper layers cause the paper to warp or even form a cylinder.

The feel of inkjet papers and photo media meets important user needs. Documents printed by inkjet should have the same surface texture, stiffness, and weight as those produced by laser printers and copiers. For consumer acceptance of photos printed by inkjet, it is important not only to deliver the image quality of conventional photographs but also to reproduce the tactile properties of the photographic surface and substrate.

4.6.2 Imaging Properties

Surface treatments and coatings, ink-receptive layer(s), and the base color of the print medium play an important role in image quality. These determine the surface gloss and finish; size, dot shape, and optical density; overall image brightness and hue; ink capacity; and dry time. Ink-receptive layers play a significant role in print fade resistance and waterfastness. The opacity and ink permeability of the substrate is important to minimize or eliminate strikethrough. Strikethrough is an undesirable property where an image printed on one side of the sheet is visible (to some extent) on the other side. This is often a problem when printing images or color graphics on plain papers, especially thinner (e.g., 16 pound) papers. Strikethrough limits the suitability of paper for duplex printing in all but the lowest quality applications. Typically, strikethrough can be minimized or eliminated by using a heavier grade of plain paper or an inkjet paper that is coated on both sides.

Substrates come in two general forms: absorbent and nonabsorbent. Paper is the most common example of an absorbent substrate; the print medium itself absorbs the ink. Because untreated paper can allow excessive penetration of ink, coatings are applied to hold the colorant near the surface to achieve high color saturation and minimize strikethrough.

Nonabsorbent substrates are used in overhead transparency (OHT) films and some photo papers to give a smooth base for transmission or reflection of light. PET (i.e., polyethylene terephthalate) is commonly used for OHTs; some photo papers use a polyester base or highly sized* papers made waterproof by laminating a polyethylene film on both sides. For these substrates, the ink receptive coating must hold the entire quantity of ink deposited during the printing process, and this can be as much as 35 ml/m^2. Although some ink vehicle may escape by evaporation over time, failure to absorb all the ink results in ink pooling on the surface, wasting the print and contamination of the printer paper path from liquid ink.

During printing, inkjet-receptive coatings control the movement of the ink vehicle on the surface of the print medium affecting color bleed, area-fill uniformity, and dot gain. Coatings remove the vehicle from the surface by absorption, and this controls short-term drying that affects smearfastness (for duplex printing) and the elimination of blocking. Blocking occurs when a sheet is stacked on top of a freshly printed sheet in the printer output tray. The printed image of the damp sheet may transfer colorant to the bottom side of the upper sheet.

Most ink-receptive coatings are composed primarily of pigments and polymers with small amounts of additives. Pigments used in coatings may be particles of silica, alumina, clay, titania, and various carbonates that are mostly non-soluble in the ink vehicle. Irregularly shaped, their dimensions can range from tens of nanometers to tens of microns, depending on the properties desired. Pigments may function as inert fillers or provide most of the ink receptivity and absorption capacity in the void space between particles.

There are two primary types of coatings: porous and encapsulating (often called "swellable").

Porous coatings are mostly silica or alumina particles held together by a polymer binder. Pore sizes range between ten and several hundreds of nanometers. The ink quickly penetrates into the voids and is captured to produce very rapid dry-to-touch times (typically less than 1 second). The void volume strictly limits the ink capacity: exceeding the ink capacity will cause liquid ink to flood the

---

* Sizing is added during paper manufacture to increase strength, water resistance, abrasion, opacity, smoothness, and optical brightness. Sizing agents include natural resins, alum, starch, waxes, and other materials.

surface. Porous coatings can deliver very high and uniform gloss, high scratch resistance, and excellent waterfastness. Because dyes are not protected as in encapsulating coatings, oxygen and other radicals can diffuse into the pores causing dyes to decompose and fade. Fade resistance is one of the major challenges in the development of porous coatings.

Encapsulating coatings use water-soluble polymers such as polyvinyl alcohol, polyvinyl pyrollidone, or gelatin. They offer complementary attributes to porous coatings: high ink capacity and good-to-excellent fade resistance but longer dry times, marginal waterfastness, and less scratch resistance. The extremely small size of dye molecules enables them to penetrate the partially solvated polymer coating. Most of the ink vehicle eventually evaporates, leaving the dyes absorbed into the polymer, which acts as an oxygen and airborne pollutant barrier. Encapsulating coatings cannot effectively encapsulate pigment-based colorants because of their larger size, and the pigments remain on the (smooth) surface forming a high-viscosity film that produces variations in gloss and often has poor smearfastness and abrasion resistance.

Polymers have a great influence on the performance and function of ink receptive coatings. For example, water-soluble or swellable polymers can be used to absorb the ink vehicle, rigid polymers can add mechanical strength, film-forming polymers can provide surface gloss, cross-linking polymers can impart durability, charged polymers can act as mordants,* and hydrophobic polymers can reduce surface tack.

Performance additives are used in small quantities (usually less than 5% by weight) to control the properties of coating materials during manufacture and adjust the performance of the finished sheet. There are numerous reasons for using additives, including defoaming, viscosity modification, leveling, surface energy modification, and control of curl.

The backside of imaging media has its own functional requirements. In the input tray, the frictional characteristics of the backside of a sheet being picked help separate it from the imaging side of the sheet below. Backside layers provide curl control so that the paper lies flat under a range of temperatures and humidities. In the output tray, the backside coating prevents printed sheets from sticking together and blocking. These characteristics are often obtained using a stacking layer formed by a water-resistant polymer with a surface textured by small plastic particles.

An off-white color and markings (e.g., manufacturer's logo, machine-readable codes, arrows, etc.) on the backside help the user load photo papers with proper side up for printing. Recently, codes have been introduced that can be read by the printer to give information about paper size, coating type, and whether the paper is correctly loaded. This improves overall ease of use by ensuring that the print mode is automatically matched to the paper and the right side is up for printing.

---

* One function of a mordant is to bind the colorant (dye) to the ink receptive layer. Inks can also contain mordants.

4.6.2.1 Plain Papers

About 98% of all inkjet printing is done on plain paper. This is often a multipurpose paper designed for copiers, laser printers, and inkjets. The surface of plain paper is usually smooth but porous. This property ensures that a multipurpose paper has good toner transfer characteristics for electrophotographic printers and copiers. Dyes can penetrate deeply into plain papers reducing color saturation and strikethrough. Because pigment particles are too large to penetrate into the paper, they stay on the surface to give the dark blacks important for printing text. Pigments typically produce duller colors than dyes on plain papers.

Plain papers are composed of fibers from hardwoods, softwoods, and recycled materials. Some have outer layers of high-quality fibers designed for surface and imaging properties. These surround a thick core of inexpensive, often recycled materials, providing bulk and strength. Sources of fiber are highly variable with season, location, and manufacturer, so additives and sizing are added to achieve consistent levels of performance for a particular brand. However, the worldwide variation in the properties of plain papers offers challenges to inkjets for the reproduction of color and sharp, black text.

Most plain papers are engineered to meet a user's primary printing needs at the lowest possible cost, about $0.01 per A/A4 sheet. These needs include good color gamut, a white background, minimal feathering, minimal cockle, reliable paper pick and minimal jamming, acceptable dry time, and good smearfastness.

4.6.2.2 Coated Inkjet Papers

Coated inkjet papers provide higher image quality while retaining some of the desirable tactile and other physical characteristics of plain paper. Papers may be coated on one or both sides. An ink-receptive coating tightly controls dot spread and dot shape and offers a bright, white background for contrast and color balance. Some coatings use fluorescent dyes to make the paper appear whiter, but these papers can have a bluish cast viewed alongside plain or bond papers in a document. To obtain high reliability in paper pick and feed, additives are used to control the surface friction characteristics.

4.6.2.3 Photo Papers

Each major inkjet printer manufacturer offers a line of photo papers to ensure the highest image quality. These use porous or encapsulating coatings over a bright-white photobase or white polymer film. Photo papers provide the greatest control over dot spread and ink penetration.

Glossy photo papers offer higher color saturation than matte papers. Glossy surfaces provide specular reflection: the light absorbed by a printed dot is reflected back at (nearly) its angle of incidence. Matte surfaces reflect light received over a wider solid angle, including room light, and colors may appear duller and less saturated because light absorbed by a dot is mixed with diffusely reflected light before reaching the eye.

4.6.2.4 Overhead Transparency Films

Transparency materials are usually coated on a single side with a swellable ink-receptive coating. To produce a reliable paper pick and ensure the ink-receptive coating is properly oriented for printing, the leading edge of the coated side of OHTs often has an adhesively backed, removable paper strip. Because the polymers used in OHT ink receptive coating have a low melting point, OHTs designed only for inkjet can melt and severely damage fuser rollers in electrophotographic (i.e., laser) printers. In recent years, the widespread adoption of PC projectors has dramatically reduced the demand for OHT films.

4.6.2.5 Specialty Media

Inkjet printer manufacturers and a large number of third-party suppliers offer specialty media for business and creative applications. These include business cards, brochure paper, textured papers, heavyweight papers for cover stock, banners, greeting cards, CD labels, and iron-on transfers.

4.7 PRINT MODES

Inkjets support a wide range of print media, and the imaging properties of different media vary dramatically in terms of dot gain, feathering, color-to-color bleed, dry time, and ink capacity. In addition, the user may at times require different levels of speed, quality, and cost per page. Plain paper, economy, and speed are important attributes for documents printed for quick review and markup. Photos and documents for external distribution usually require the highest levels of print quality and materials. To meet these needs, inkjet printers offer flexible solutions called print modes.

Inkjet printers almost never scan the printheads across the page and print a dot at every required location in one pass. Instead, they can print in single, multiple, unidirectional, and bi-directional passes over the paper depositing ink drops in patterns and combinations of bewildering complexity that are, fortunately, controlled by printer firmware from simple choices made by the user.

Selection of the print mode and paper type is made in the print driver, which pops up whenever the user prints a page or document. Figure 4.11 shows an example of quality choices among draft, normal, and best print modes and media choices among a variety of paper types.

Draft mode uses the least amount of ink and is the fastest and most economical print mode. It is usually reserved for plain paper and locked out when the user selects expensive media, such as photo papers. In draft mode, ink use is reduced by not printing every dot. This is called "dot depletion," and it helps the sheet to dry quickly and prevents dots from merging when printing in a single pass. Solid colors will appear de-saturated because the paper background is not completely covered with dots. To obtain the highest possible print speed, the print carriage scans only once over the print zone and the paper is advanced the full

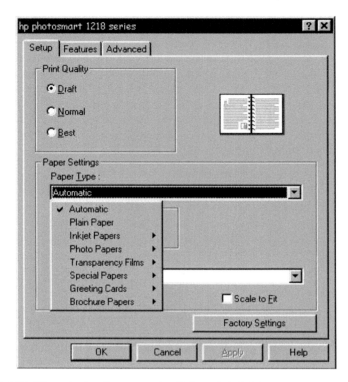

FIGURE 4.11 Magenta density ramp using light and dark inks.

swath height of the printheads. The black printhead often has a wider swath (and more nozzles) than the color printheads on low-cost printers because of the need for fast text printing. In this case, black and color scans may be made separately. One of the common issues in draft mode is that nozzles that are clogged or produce misdirected drops will leave a white (or uncolored) band that the eye is very adept at seeing.

In some printers, draft mode is printed bi-directionally producing color shifts between bands that depend on the direction of carriage travel. When multiple color dots overprint, the hue depends on drop order. For example, a different blue is obtained by printing cyan first and then magenta compared to magenta followed by cyan. This effect is called bi-di hue shift. Printing bi-di can also cause vertical lines to be misaligned between swaths when drop placement characteristics depend on the direction of printhead travel and cannot be eliminated by compensating the timing of drop ejection. Some printers can print and scan a calibration page with test patterns to align colors, minimize bi-di misalignment (usually only for normal and best modes), and minimize paper advance banding.

Normal mode provides a good balance between speed and quality for most applications. Here, solid colors are printed with full coverage and text-enhancing processes, such as black fortification, are used. Partial swath advance of the paper allows multiple scans of the printheads over the same area so that good nozzles substitute for unfit nozzles to print all pixel rows. Small errors in drop trajectories, nozzle-to-nozzle drop volume variations, and paper advance are effectively hidden by multipass printing in normal mode.

Normal mode is almost always printed uni-directionally to eliminate bi-di hue shift and drop placement errors. However, some high-speed workgroup inkjet printers offer fast normal print modes with correction algorithms that minimize bi-di color shifts.

Best mode is the slowest and highest quality mode because more passes are used over a given area to average out nozzle-to-nozzle variations, and ink has the longest time to be absorbed by the print medium, minimizing color-to-color bleed. In best mode, every neighbor to any given dot will be printed by a different nozzle on a different pass. Best mode is typically the default for printing photos on photo media, but recent advances in media and printhead design allow some printers to produce high-quality 4" × 6" photos in about 15 seconds in a fast draft mode that are nearly indistinguishable from normal and best modes.

The number of ink droplets printed into each pixel depends on the quality mode and the ink capacity of the print medium. Proper color rendering depends on how the print medium's ink capacity and dot formation processes affect tone reproduction characteristics. This is why selecting the correct medium is important to meeting print quality expectations. For example, selecting best mode and glossy photo paper but loading plain paper in the input tray will put too much ink on the paper, giving a muddy, dark image with poor color fidelity.

Selecting the proper print mode and print medium has been a source of user confusion with desktop inkjet color printers. In recognition of meeting a user need for reliable imaging, some manufacturers have added an automatic media type sensor to the printer. This sensor matches the print mode to the type of medium sensed in the print zone. These sensors use light emitting diodes (LED) and a photodetector to measure the difference between specular and diffuse reflection from the surface of the print medium, and they can distinguish between plain paper, coated paper, photo paper, and overhead transparency film. Printed, machine-readable marks on the backside of photo papers can further refine the selection to optimize photo print modes to different photo coatings.

4.8 HIGH-FIDELITY COLOR

4.8.1 OVERVIEW

Desktop color inkjet printers place individual colored dots a few tens of microns in diameter on addressable grids as fine as 5760 points per inch. A four-color CMYK* printer uses cyan, magenta, yellow, and black inks chosen to deliver a large color gamut with highly saturated colors, dense shadows in images, and high-contrast black text. The yellow ink adds little optical density and is used to

---

* This is the common designation for a cyan (C), magenta (M), yellow (Y), and black (K) ink system.

control the hue of the printed pixel. Patterns of isolated cyan, magenta, and black dots in image highlights and midtones can have high contrast against the bright white background of inkjet photo papers. This produces a variation in lightness perceived as an undesirable, random texture called grain. Minimizing the visible grain in an image is a key objective for achieving high image quality.

Although individual printed dots span a very small visual angle at normal viewing distances, the contrast sensitivity of the human visual system allows lightness variations to be perceived.[4] However, the eye is less sensitive to color variations under these conditions. The first implication of all this is that very tiny dots of full-density cyan, magenta, and black inks are required before they can produce images free of noticeable grain. This requires small drop volumes, usually about 2 pl or less. The second implication is that combinations of cyan, magenta, yellow, and black dots can produce the perception of millions of distinct colors, and this principle forms the basis for all color halftoning algorithms.

A binary pixel is one in which a single drop of ink fills the pixel area. In a CMYK system, binary pixels offer only eight addressable colors: white (background), the four primaries (cyan, magenta, yellow, and black), and the three secondaries (red, green, and blue). Even at 600 pixels per inch, the granularity of binary pixels printed with CMYK inks is too high for true photographic quality at the normal viewing distance of a handheld print. Some large-format CMYK inkjets use a 600 dpi binary print system. Unlike typical photographs, these large (poster-size) prints are viewed from a distance of several meters where the visual angle subtended by the pixels is equivalent to more than 2400 dpi in a handheld print.

Reductions in granularity have historically been achieved with smaller dots. However, the trend in Figure 4.2 shows diminishing reductions in dot size with drop volume below about 4 pl. Also, 2 pl may be close to a practical manufacturing limit for drop-on-demand drop generators because features with very small physical dimensions are required. Small features will be more susceptible to clogging and sensitive to manufacturing tolerances. Most manufacturers are now producing the smallest drops (2–5 pl) that can be reliably ejected for their dark inks.

The most effective and practical means to reduce granularity uses a halftone pixel. In a halftone pixel, the printer has control of the dot size, number and placement of dots within a single pixel, dot optical density, or all three. The method used by most desktop color printers is a combination of small drop volumes and "light" primaries, typically light cyan (c) and light magenta (m). Such a six-color system is called a CcMmYK printer, and processes using more than four primary colors are known generically as high-fidelity color systems. This section will show how such color systems can reduce image granularity and improve the printer's color gamut.

Recently, desktop color printers have introduced systems with one or more neutral gray inks. This offers several benefits: a reduction in image grain, improved dark colors and shadow detail, better and more stable neutral tones, and reduced metamerism. When composite grays are printed from combinations of cyan, magenta, and yellow inks, any variation in drop volume from one of the primaries can cause the gray to shift toward red, green, or blue. Printing neutral tones directly with gray inks avoids this problem.

Color constancy effects are particularly noticeable in gray tones and in black and white prints. Composite grays usually exhibit some degree of metamerism: a gray viewed under one illumination (for example, tungsten light) can appear greenish or reddish under daylight or florescent light. Gray inks with uniform spectral reflection can significantly reduce metamerism.

Multi-ink strategies can be used to increase gamut size by adding primaries around the gamut equator: highly saturated oranges, greens, or blues can significantly increase gamut volume. Use of additional colors has also been considered to improve the spectral match for reproduction of fine art prints with less metamerism, and can bring more Pantone® colors in-gamut for graphic design applications.

4.8.2 COLOR SEPARATION

In CMYK printing, it is well known that many colors can be produced from different ink combinations, especially different amounts of black relative to cyan, magenta, and yellow. In mathematical terms, this is an under-constrained situation: tristimulus human color perception is well modeled with three dimensions while four inks are often available to reproduce color. There are a wide variety of undercolor removal (UCR) and gray component replacement (GCR) strategies to determine how much black ink to use in the color separation. This produces a more neutral, higher-density black and reduces the amount of ink used compared to composite (CMY) black.[5] Factors to be considered include image grain, ink usage, ink capacity of the print medium, dry time, and color constancy.

Adding colors, such as light primaries and grays to a CMY system permits printing more colors per pixel. With CcMmY inks and three gray inks (light, medium, and dark), the number of theoretical dot combinations per pixel gives over 72 million addressable colors.[*] While all these colors may not be practically distinguishable, and printer color tables will use a subset of all the possible combinations, this is continuous-tone printing for all practical purposes.

In subtractive color printing, the base color is the white paper and care is taken in photographic media to provide a bright, neutral background. In the light tones of image highlights, light cyan and light magenta inks are printed with yellow on this white background.[**] There is low contrast dot-to-dot and for the dots against the background. With increasing print density, the light inks are printed, with area coverage increasing up to 100%. Because this produces only a midtone color, higher density is achieved by printing full-density (i.e., dark) cyan, magenta, and black inks on a background of light inks. This serves to reduce

---

[*] This is HP's claim for its PhotoREt Pro color layering technology, where up to 29 drops of ink can be printed in a single 300 dpi pixel.

[**] Because the full-density yellow ink has very low contrast on the white background, there is little justification from enhanced image quality for adding a light yellow printer. The additional nozzles and drive electronics, which add to printer cost, are better distributed among the CcMmYK primaries to obtain higher color throughput.

the contrast between adjacent dots for lower perceived image grain. This process is shown schematically in Figure 4.12 for a magenta density ramp. Similar ramps are used for light and dark cyan and combinations of gray inks.

In this figure, an output level of magenta is produced by combination of light and dark components. For low magenta densities, only the light ink is used. At the midtones, dark magenta is introduced on the light magenta background. At high magenta densities, mostly dark magenta is printed and the amount of light magenta is reduced to avoid exceeding the ink capacity of the print medium. As is typical of all color separation schemes, the quantity of each ink is varied continuously to produce a smooth density ramp.

4.8.3 PRINTING WITH LIGHT INKS

4.8.3.1 Light Ink Strategies: Halftoning

The digital halftoning process plays an important role in reducing image grain in CcMmYK printing because it accounts for dot-to-dot contrast information. The image processing pipeline for digital color printers, including those that use light inks, is shown schematically in Figure 4.13.

The continuous tone image data from the source image in its color space (e.g., RGB, sRGB, and CIELAB) is first transformed into printer CMY densities, considering the spectral characteristics of its inks. This is called the device (color) space. At this point, input colors that are outside the printer's color gamut are brought into the gamut using various strategies. Next, the printer CMY densities are transformed into ink amounts, taking into account drop volumes, dot size on a specific print medium, the type of print medium and its ink capacity, and the number and spectral absorption characteristics of the inks. Additional factors may be considered; for example, dye-based black ink can be mixed with dye-based colored inks, but pigment black inks are not generally mixed with dyes in images on photo media. Given the amount of data to be processed in real-time while printing, these color transformations are usually done in a lookup table (LUT) in the printer's firmware.

Finally, halftone processing determines the number and location in each pixel for dots of each ink color. The halftoning process typically takes CMYK densities for each pixel and computes the number and location of individual ink dots. At the level of individual pixels, halftoning may introduce some error between the desired and printed color, particularly when only three or four inks are available. Some algorithms, such as error diffusion, operate by distributing this error to neighboring pixels and the spatial response function of the eye produces the perception of a uniform, desired value. Error diffusion becomes less important when more addressable colors are available (e.g., with six- and eight-ink systems) because the color errors are much smaller.

A color plane is an array of density values for a single color at each pixel in the image. In a CMYK printer, there will be three color planes: C, M, and Y, because black pixels can be represented by C + M + Y. The halftoning process converts the

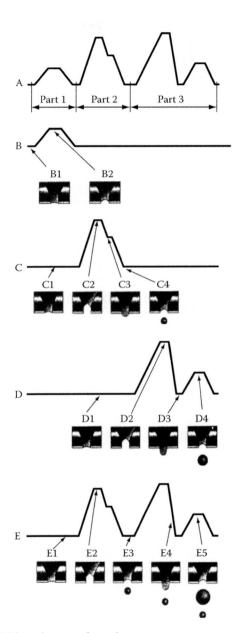

FIGURE 4.12 An inkjet printer configuration menu.

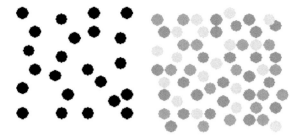

FIGURE 4.13 (See color insert following page 176.) Image-processing pipeline.

three color planes into four dot masks, which are instructions to the printheads to print or not to print a dot at each possible location on the image. There will be six dot masks in a CcMmYK printer. Halftoning can operate on each color plane independently or all color planes together. If the colors are halftoned independently of one another, there is no control over how dots of one color will spatially interact with dots of another color; they may partially or completely overlap, producing unintended results such as dark dots. If the color planes are halftoned together, then the dot patterns for each color can be interdependent and control can be exercised over dot overlap and adjacency to produce optimal results.

There is a significant advantage to exercising interdependent color dot placement. In many cases, image grain can be reduced by preventing dot overlap until midtones or darker colors are printed. In the case of independent halftoning, light cyan and light magenta dots may overlap, producing a bluish dot that is darker than either the light magenta or the light cyan individually. In the case of a 20% fill factor for both inks, c and m will overlap at about $0.2 \times 0.2 = 0.04$ or 4% of the print locations. Yellow and the light inks interact similarly. Statistically, all three inks will overlap at some points, producing a gray print spot that is darker than any of the three primary inks used. The overlapping dots will result in more contrast between adjacent print locations (bluish, greenish, reddish, or gray against the white background) than any single dot alone. The resulting print appears grainier than when the dots can be distributed to minimize overlapping (i.e., interdependent halftoning). Of course, once the required density from one or more colors exceeds 100%, some locations must be overprinted. However, even for darker colors, some reduction in grain can be realized by preventing yellow and dark cyan and dark magenta from printing in the same location (making the blackest possible dot) until necessary.

4.8.3.2 Light Ink Strategies: Light Ink Color

An obvious method for producing the light inks is to simply dilute the colorant used in the dark inks with more (clear) ink vehicle. This produces a less chromatic ink of the same hue as the dark ink, and light inks are sometimes made this way. An advantage of dilution is that the light and dark inks will have similar, but scaled, reflectance spectra. This means that they will change color similarly under different light sources; they exhibit similar color constancy. Thus, a range of gray values from white to black (i.e., a neutral ramp) produced with light inks for highlights, a mixture of light and dark inks for the midtones, and dark inks for shadows will shift in the same way as the image is viewed under different illumination sources. An additional benefit of hue matching between light and dark primaries is hue constancy in the midtones as dark inks replace light inks.

For example, with similar color constancy between light and dark primaries, if the hue shift observed between two different illuminants is toward red, then all tones from light to dark along the neutral ramp will red shift. If the light inks and dark inks are formulated without similar color constancy, then a density ramp appearing neutral when viewed under one illuminant may shift to green for some gray values and to red for others when viewed under another illuminant. This wandering neutral poses significant image quality problems and can cause a color printer to be essentially useless for printing high-quality black and white photos.

A potential problem in making light primaries by dilution is unacceptable fade resistance for the light inks. Thus, manufacturers may use more fade-resistant colorants for the light inks with an effort to ensure that the light and dark inks exhibit similar hues viewed under the most important illumination sources.

The ink capacity of the print medium must also be considered when designing the light inks. If the optical density is too low, excessive amounts of the light ink may be required to achieve midtone densities, and this could exceed the ink capacity of the print medium, especially for porous coatings.

4.8.4 PRINTING WITH DARK INKS

Printing with more than four color primaries is common in commercial offset lithography, digital presses based on liquid toner electrophotography, and large format inkjet printers. Additional inks permit direct printing of spot, logo, and Pantone® colors used for corporate and brand identities. They also increase the gamut size, improve color constancy, reduce ink usage, and create smooth, grain-free gradations for certain critical colors such as skin tones. As of this writing, desktop color printers offer up to nine inks adding to CcMmY various combinations of gray, blue, red, and green inks.

Achieving a larger printed color gamut is the usual reason for adding primary colors. Because of the large installed base of six-color lithographic presses, much of the effort in this area has been on CMYK systems with two additional primary inks.

4.8.4.1 Dark Ink Strategies: Selecting Additional Primaries

Consider now how green and blue inks can expand the gamut of a printer that complies with the SWOP* color standard. Being able to print the full SWOP gamut (as well as other standards for commercial color printing used in Europe and Japan) is important for accurate prepress proofing on inkjet printers. This

---

*For example, Standard Web Offset Press specification: ISO 2846-1, Graphic Technology - Color and Transparency of Ink Sets for Four-Color Printing - Part 1: Sheet-fed and Heat-set Web Offset Lithographic Printing (1997).

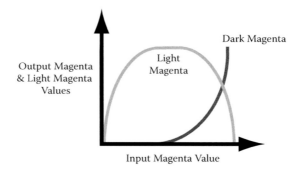

FIGURE 4.14 Reflectance of SWOP primaries.

example is only for tutorial purposes and is not provided as an ideal implementation. It has been highly simplified to reduce the complexity of the analysis while
preserving an indication of the potential benefits.

Figure 4.14 shows measured reflectance curves for a SWOP press. Individual
reflectance curves are presented for the substrate (i.e., white paper) and the cyan,
magenta and yellow colorants.

Figure 4.15 shows the measured reflectance of two secondary colors, green,
and blue, produced by overprinting the primaries. The figure also presents scaled
curves produced by hypothetical green (G) and blue (B) primaries for use with
the substrate and CMY colorants of Figure 4.14. The curves for G and B are not
based on any specific colorants; they have been chosen to yield higher chroma
at roughly the same hue angles of actual SWOP green and blue secondaries. This
also reduces the amount of cyan and yellow inks to print green, cyan, and magenta
inks to print blue.

Figure 4.16 presents gamut equators in CIELAB color space, where the x-
axis is a* (red-green) and the y-axis is b* (yellow-blue). The gamut of the CMY
inks of Figure 4.14 is shown as a solid line. The gamut of the CMYGB system

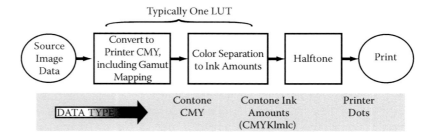

FIGURE 4.15 Reflectance of overprinted secondaries and G, B inks.

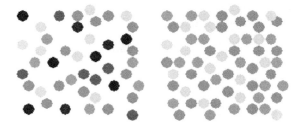

FIGURE 4.16 (See color insert following page 176.) Gamut of CMY and CMYGB printers.

is shown by the dotted line. Here, G and B values from Figure 4.16 have simply been substituted for the composite green and blue, although Viggiano[6] describes how secondaries might be predicted from a given set of primaries.

The original CMYK printer had a gamut volume of roughly 415,000 in CIELAB color space. Adding the green printer (G) increases the CMYG gamut volume to 498,000; adding the blue printer increases the CMYB gamut volume to 518,000. The CMYGB printer delivers a gamut volume of 601,000.

The hypothetical CMYKGB printer produces a gamut volume about 40% larger than obtainable with CMYK primaries. Because the blue and green colorants are hypothetical, this gamut expansion is not necessarily typical. Inkjet printers using cyan, magenta, and yellow inks with higher chroma compared to SWOP would not realize such large increases in gamut volume by the addition of blue or green printers. In this case, additional primaries such as violet or orange are often used to achieve gamut expansion.

4.8.4.2 Dark Ink Strategies: Image Processing

The image processing used with more than four full-density primaries is similar to the flowchart shown in Figure 4.13. In this case, the color separation process between printer CMY conversion and halftoning involves the custom colorants instead of light cyan and magenta. There are many to perform this separation, but one example can be demonstrated that is similar to the gray component replacement process.

For a six-color printer with additional green and blue primaries, a hue sequence determines the nearest color neighbors. For example, printing cyan and yellow primaries produces green, and printing cyan and magenta produces blue. The hue of the secondary color (i.e., green and blue) will not necessarily be a simple linear mixture of the two, but it generally falls somewhere between the hues of the two primaries.

Assuming a gray component replacement (GCR) algorithm has been applied to the color data to control the black printer, the values cyan', magenta', and yellow' in the following example are the remaining colors required to produce pixel chroma. For 8-bit color planes, these values are typically expressed as integers ranging between 0 and 255.

A supplementary color replacement (SCR) algorithm can be applied to the color data after GCR. For green, a simple SCR process is:

$$green = min\ (cyan',\ yellow') \tag{4.1}$$

$$cyan'' = cyan' - green \tag{4.2}$$

$$yellow'' = yellow' - green \tag{4.3}$$

where cyan″ and yellow″ are the new printer cyan and yellow values for the printer given any value of green produced by Equation 4.1.

In the same manner, a simple blue SCR is:

$$blue = min\ (cyan',\ magenta') \tag{4.4}$$

$$cyan'' = cyan' - blue \tag{4.5}$$

$$magenta'' = magenta' - blue \tag{4.6}$$

where cyan″ and magenta″ are the cyan and magenta values for the printer given any value of blue from Equation 4.4. It is assumed, of course, that blue and green do not overprint in any pixel.

Note that Equation 4.1 through Equation 4.6 are highly simplified and could add granularity where the minimum of cyan, magenta, and yellow is not zero. This recalls the earlier discussion of dot visibility as a function of color and density. However, these equations provide an example demonstrating how device-to-device conversions might be performed during the color separation process with four full-density color primaries. Additional and more complex separation schemes may be found in the literature.[7]

4.9 CLOSURE

Like the printed word centuries ago, the production of color images and documents was once the exclusive domain of a small number of highly trained and skilled people working within narrow professional communities. The past 2 decades of extraordinary developments in desktop color inkjet technology and inkjet printer features have enabled a level of personal expression through word and image scarcely imaginable by either Gutenberg or Lord Rayleigh.

REFERENCES

1. T. Kitahara, Ink-jet head with multi-layer piezoelectric actuator, *Proc. IS&T 11th Int'l Congress on Advances in Non-Impact Printing Technologies*, IS&T, Springfield, VA, 1997, pp. 346–349.

2. M. Usui, Development of the New MACH (MACH with MLChips), *Proc. IS&T 12th Int'l Congress on Advances in Non-Impact Printing Technologies*, IS&T, Springfield, VA, 1998, pp. 50–53.

3. H. P. Le, Progress and trends in ink-jet printing technology, *Journal of Imaging Science and Technology*, 42(1), pp. 49–62 (1998), and on the IS&T website: www.imaging.org.

4. Lau and Arce, Modern Digital Halftoning, p. 42 (2001).

5. M. Southworth, *Color Separation on the Desktop*, Graphic Arts Publishing, Livonia, NY (1993).

6. Viggiano, Colorant Selection for Six-Color Lithographic Print, 6th CIC (1998).

7. E. Stollnitz, Reproducing Color Images Using Custom Inks, Ph.D. Thesis, 1998. (http://grail.cs.washington.edu/theses/).

5 Laser Printer

Fumio Nakaya and Yasuji Fukase

CONTENTS

5.1 HISTORY

Electronic image information is converted into the hard print that human eyes
can see; the equipment using an electrophotographic method with a laser beam
as a light source is called a laser printer. An electrostatic latent image is formed
on a photoreceptor by sequential laser spots that are modulated based on the
image signal with a rotated polygon mirror, and a visualized image is formed by
the electrophotographic method.

Electrophotography started from the copying machine application; it has
evolved for years and now can handle digital signal information by using a laser
beam source. The laser beam printer with a He–Ne laser was put into practical
use for the first time as a high-speed output terminal of a computer at IBM
Corporation. Now it serves an important role as an output device for workstations
and personal computers. Furthermore, color copying machines and printers are
advancing by using the electrophotograph method.

As another effect of being able to deal with a digital image signal, electrophotography has affected progress of this colorization technology. The raw characteristic of electrophotography is very high contrast reproduction, and smooth
gradation is difficult to reproduce. The digital image signal is compensated on
image-processing technology, and reproduction characteristic of an electronic
photograph has become possible with the linear input/output characteristic of
image density.

Moreover, rapid development of electronic information machines and equipment has a great influence in all areas, including intellectual activity and production activity, and recombination of industrial structure is progressing steadily.
The conventional division-of-work organization in printing from individual work
to the printing industry is about to collapse.

Enhancement of network infrastructures and improvement in performance
has developed a new print system called Over the Internet Printing or Distribute
and Print. Under this new infrastructure, it has researched what is the most
balanced (quality/cost) total system of printing through printing production work
to delivery to the customer.

Laser printers have extended the range of print speed from 1 ppm (page per
minute) for personal usage to 180 ppm for on-demand printing. Moreover, a color
laser printer also has the capability of producing several ppm to over 110 ppm
and has a key position in the print market.

The improvement in image quality by digital image data processing is remarkable, not only with electrophotography but also with various marking technologies, such as inkjet, dye sublimation, silver halide, and conventional lithographic
printing. The laser printer with the electrophotographic system is in a middle
position for quality of image and productivity in these hard copy printing systems.
In the future, the laser printer will possibly extend the application range to both
ends, but it also runs the risk of shrinking the range by the progress of other
marking technologies.

The printer market overall had about 90 million unit shipments annually in fiscal year 2001, and inkjet systems, which show remarkable growth in home use applications, constituted 70% of this market; whereas electrophotographic printers constituted only about 12%. However, from a revenue perspective, inkjet systems constituted about $10 billion and electrophotographic laser printers, including monochrome and color printers, constituted $13 billion. Moreover, consumable sales, such as ink for inkjet printers or toner for laser printers, matched hardware sales, with inkjet cartridges at $8.5 billion and electrophotographic cartridges at $10 billion in 1998.[1]

Regarding color printer products, most inkjet printers are color printers. However, shipment of electrophotographic color printers is still small, with 166,000 shipments in 2001 in the Japanese market. (This is equivalent to 14% of the number of monochrome page printers shipped in Japan.)

1996 print volume data show 210 billion prints in Japan and 600 billion prints in United States. Print volume is increasing rapidly vs. copy volume. Now the number of prints is close to or exceeds the number of copies, and the trend is expected to continue.

5.2 MARKING TECHNOLOGY

5.2.1 ELECTROPHOTOGRAPHY

The Carlson process, developed in 1938, is one of the most common electrophotographic methods. The basic principle of the Carlson process is charging photoconductive material uniformly, exposing an optical image to form an electrostatic latent image, developing a latent image with particles that have an electrostatic attraction and visualizing the image, transferring the particles onto media, and fusing the particles permanently onto the media.

A detailed explanation of the electrophotographic process is available in technical books.[2,3] The electrophotographic process, as shown in Figure 5.1, iterates the seven steps of charging, exposing, developing, transferring, fusing, discharging, and cleaning consecutively to get multiple sheets of images. The electrophotographic process forms a latent image each time, so it is also called no-plate printing.

For color print, the above electrophotographic process is iterated four times for cyan, magenta, yellow, and black images, and then the colors are overlapped. Color electrophotographic system architectures are categorized into three types: (1) overprint on photoreceptors, (2) overprint on intermediate media, and (3) overprint on paper[4] (see Figure 5.2).

Other categorizations can be done to count the number of photoreceptors in a system. A single photoreceptor type with multiple development housing is called a multi-path type, and multiple photoreceptors, each with one development housing, is called a tandem type. Figure 5.3 shows a typical tandem structure.

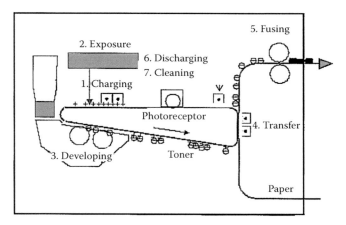

FIGURE 5.1 Electrophotography process.

Table 5.1 shows a comparison of various types of electrophotographic system architectures. Tandem type is theoretically four times faster than multi-path type; however, mechanical architecture is more complicated and a sophisticated color-registration mechanism is required. The tandem type system was first introduced in the high-end and high-price market. Another feature of the tandem type system architecture is to handle a wider range of media due to its simple media path, and it can be dominant with an evolution of a low-cost color-registration mechanism.

5.2.2 MARKING PROCESS

A schematic diagram of a multi-path intermediate belt transfer color laser printer is shown in Figure 5.4.[5] Yellow, magenta, cyan, and black image signals are converted from device-independent signals to device-dependent signals, in terms of color rendering, Modulation Transfer Function (MTF), Tone Reproduction Curve (TRC), halftone screening, etc., and sent to a laser Raster Output Scanner (ROS) driver. The laser ROS exposes the photoreceptor to discharge the photoreceptor electrostatic charge and form a latent image of the original. The development device, in this case a rotary development housing, develops the latent image on the photoreceptor with a toner, by electrostatic force. Each color is transferred to the intermediate transfer belt, color by color, and four toner layers are transferred to media. Finally, the four-color image is melted onto the media by a fusing device and fixed at room temperature. The rest of the toner on the photoreceptor is removed with an elastomer blade to be ready for the next cycle. In the case of black, the printer development device is stationary and the developed toner is directly transferred from the photoreceptor to the media.

A schematic diagram of a tandem intermediate belt transfer color laser printer is shown in Figure 5.5.[6]

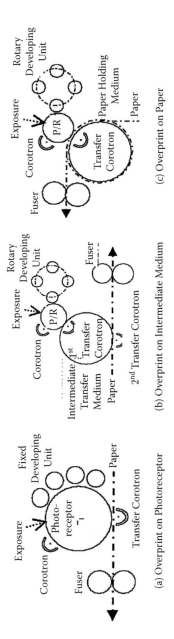

FIGURE 5.2 Multi-path color laser printer.

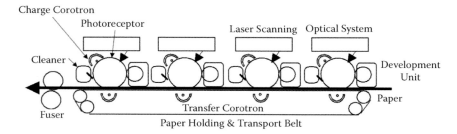

FIGURE 5.3 Tandem color laser printer (overprint on paper type).

TABLE 5.1
Comparison of Various Types of Color Electrophotography Architecture

	System Architecture	Performance
Multi-Path	Overprint on photoreceptor	Few composition element parts of a mechanism.
		Few color-to-color registration errors.
		Latent image is influenced by the last development image.
Multi-Path	Overprint on intermediate medium	Possible to handle wide range of media.
		Image noise is accumulated through twice-transfer process (toner disturbance at transfer area).
Multi-Path	Overprint on paper	Stable latent and developed image
		The kind of applicable media is restricted, for it needs to wrap onto transfer drum.
Tandem		Same productivity between mono color and color mode.
		Factors of color registration errors increase.
		Increase composition element parts and unit cost.

5.2.3 TECHNOLOGY ELEMENTS

5.2.3.1 Photoreceptor

An organic photo conductor (OPC) is commonly used as a photoreceptor. The OPC has a layered structure consisting of an aluminum base plate, an electrical charge blocking layer (BL), an electrical charge generation layer (CGL)m and an electrical charge transport layer (CTL). Key factors in the OPC are

- Interference of incident light
- Electrical charge uniformity along its surface
- Electrical charge cyclic stability
- Photoreceptor drum eccentricity

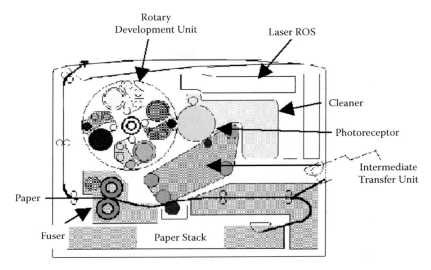

FIGURE 5.4 (See color insert following page 176.) A schematic diagram of multi-path intermediate belt transfer color laser printer.

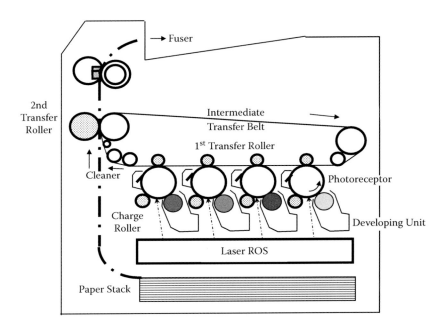

FIGURE 5.5 (See color insert following page 176.) A schematic diagram of tandem intermediate belt transfer color laser printer.

The photoreceptor drum eccentricity, electrical charge uniformity, and cyclic stability normally should be less than 50 microns and 20V.

5.2.3.2 Charging

A charging device puts an electrical charge uniformly onto the photoreceptor. The charging sign depends on the photoreceptor and toner material. Various types of charging devices are available, such as wire type, pin type, and roller type.

The wire type is called a corotron. It has a fine wire (diameter: 50–60 micron meters) and a nearby grounded shield. When a high voltage (5–10 kV) is applied between the wire and the shield, the air near the shield will become ionized. The pin and roller types are almost the same as wire except for their shapes and distances. Key factors in the charge device are geometrically skewed against the photoreceptor and dirt management to prevent its contamination.

5.2.3.3 Laser ROS Exposure

The laser ROS exposes the photoreceptor to discharge the photoreceptor electrostatic charge and form a latent image of the original. Figure 5.6 shows an example of laser ROS optics. Quantized image signals are modulated for tone reproduction and sent to the laser driver circuit; then a laser beam is emitted, with the beam hitting a polygon mirror, reflecting from it, focusing on a photoreceptor surface as a spot, and scanning the photoreceptor surface diagonal to the process direction.

There are two types of writing methods. One is to write white, which exposes the background area of the original (similar to negative in photography), and another is to write black, which exposes the image area of the original (similar to positive in photography). The latter case is more popular. Key factors in laser ROS optics are to have suitable laser power with photoreceptor photosensitivity, laser spot size, and beam positioning along the process direction. Only a few microns of beam disposition may cause banding defects such as a dark and light stripe along the perpendicular to the process direction. Also, laser ROS jitter and wobble may cause graininess, so those performances should be better for color printers than for black and white.

A laser diode emission wavelength of about 760 nm–830 nm, which is similar to application for a compact disc reader, is normally used.

A smaller laser ROS beam spot size gives higher image quality. The laser ROS beam spot size is proportional to the laser diode emission wavelength.[7] Therefore, 670 nm (GaAlP; used in DVD) or 400 nm (GaN) laser diodes are candidates for the next laser ROS.

There are two methods modulating laser ROS intensity. One is emission intensity modulation, and the other is emission pulse width modulation. In either case, it is important to keep the linearity between the image signal level and the laser beam intensity. Basically, some compensating circuit for temperature dependency of the laser light source is needed. Additionally, it must treat carefully any response delay in laser pulse rise time. Figure 5.7 shows an

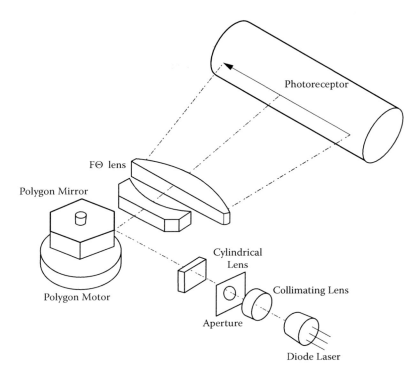

FO lens

Polygon Mirror

Polygon Motor

Photoreceptor

Cylindrical
Lens

Collimating Lens

Aperture

Diode Laser

FIGURE 5.6 Laser ROS optics.

example of nonlinearity in pulse width modulation.[8] It needs some artifice to compensate for it. For example, in Figure 5.7, halftone image density becomes stable when active pixels are set side by side, compared with dispersed layout of the active pixel case in the ordered dither screen method. This filling order helps to stabilize electrostatic contrast in electrophotography.

For faster laser printers, various multiple laser beam approaches have been commercialized or are under development as shown in Figure 5.8.[9] All approaches have a multiple laser beam along the process direction to gain higher writing speed. A two-dimensional laser array has been proposed as a static scanning device to eliminate mechanical scanning devices such as polygon mirror systems.

5.2.3.4 Development and Developer

The development device develops a latent image on the photoreceptor with powder ink, called toner, by electrostatic force. The toner is carried to the photo-receptor by a carrier, which is oppositely charged during the mixing process in the development housing. The toner and carrier mixed together is called the

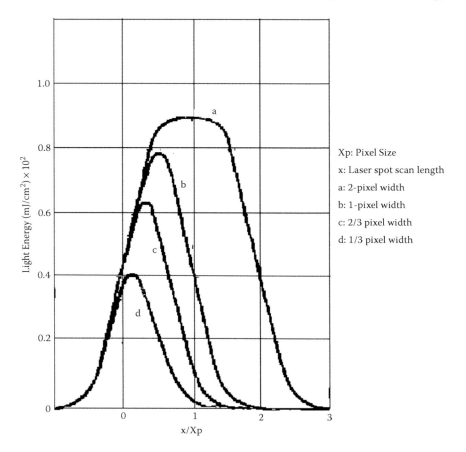

FIGURE 5.7 Nonlinearity of light energy in pulse width modulation.

developer. This process is done four times before proceeding to the next process or one color at a time.

The primary subjects for designing the developing unit for color printers are the layout of four or more color development devices along the photoreceptor, and the method of supplying a significantly higher amount of toner into the development device.

Commercialized development device layouts, for example, include rotary (CLC-1, CLC-500, A-color), horizontal (CLC-200), and vertical (CF-70). All of these layouts move each color development device to switch colors during the marking process. Thus, special care is needed to minimize mechanical vibration caused by this movement to prevent the banding defect.

The colored original image area is typically five times greater than black and white originals, and toner consumption is higher and faster. Thus, better toner and carrier mixing characteristics, called admix characteristics, are required.

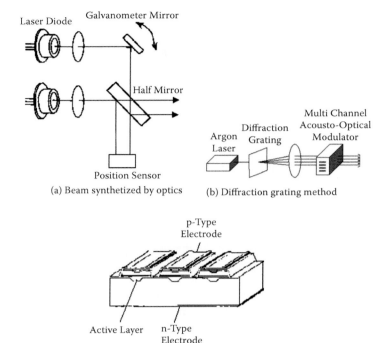

(a) Beam synthetized by optics   (b) Diffraction grating method

(c) Beam emission diode array method

FIGURE 5.8 Multiple laser beam approaches.

The most significant aspect of achieving higher image quality in electrophotography using powder toner is to make the toner particle size as small as possible. For color toner, good pigment dispersion in resin provides a more desirable density.

Development methods are generally classified into two types. One is powder toner development, as shown in Table 5.2, and the other is liquid toner development, such as pigment-dispersed iso-paraffin. The typical two-component magnetic brush development method and magnetic single-component development method are shown in Figure 5.9.[10]

In the two-component magnetic brush development method, the carrier has three functions: put an adequate electric charge quantity to the toner by friction charging, carry the toner to the development zone, and remove any unwanted toner from the photoreceptor. However, due to life issues (the deterioration of triboelectric charging performance with toner contamination on carrier surface) and complicated mechanisms, magnetic single-component development is dominant, especially in black and white printers. For color, due to magnetized toner opaqueness, magnetic single-component development is not suitable and two-component magnetic brush development method is used. The non-magnetic toner development method has good color reproduction characteristics and is

TABLE 5.2
Electrophotographic Development Methods

Method	Toner	Carrier	Process
Cascade	Colored pigment main material: thermoplastic resin	Glass bead, steel ball, etc.	Developer (mixture of toner and carrier) is sprinkled over electrostatic latent image using gravity, and only toner is made to stick to photoreceptor.
Two component magnetic brush development	Colored pigment main material: thermoplastic resin	Magnetic particle: steel powder, ferrite, etc.	Developer is adhered to a cylinder surface with magnetic force. Developer is made into the shape of a brush. Electrostatic latent image is rubbed by the brush, and toner is transferred to a latent image on a photoreceptor.
Non-magnetic single component development	Colored pigment main material: thermoplastic resin	No need	Toner is contacted with the cylinder sleeve (conductive rubber or resin tube), conveyed by image force, and transferred to latent image on photoreceptor.
Magnetic single component development	Colored pigment thermoplastic resin and magnetic particle	No need	Toner is held on the cylinder surface with the magnet inside it, and transferred to latent image on photoreceptor.

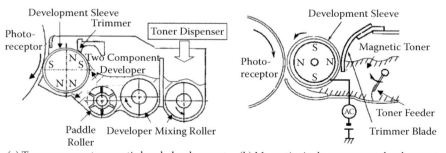

(a) Two component magnetic brush development   (b) Magnetic single component development

FIGURE 5.9 Typical development method.

smaller and simpler than the two-component magnetic brush development method. However, it has several issues such as the stability of frictional electrification, the uniformity of the thin toner layer, and the maintenance of the thin gap (within about 200 microns) between the photoreceptor and the development sleeve.

5.2.3.5 Transfer

Developed toner on a photoreceptor is transferred to either the final media to form a hard copy of the original image or to an intermediate transfer material once and then transferred to the final media. Generally, in this process, an electrostatic force is applied. Corona transfer and roller transfer are popularly used. For color print, this process is done either once (four colors at once) or four times (one color at a time).

Difficulties in this process are to achieve a good color-registration level and to achieve a good transfer ratio through the four-toner layer. The requirement of the color-registration error level depends on the original and the type of halftone screen used. Normally about 100–150 microns is good enough. The stress case is a uniform gray with the same halftone screen angle. For example, a 200-line-per-inch screen case, 64 microns out of the color-registration error causes severe color shift. It can be resolved by adopting a different halftone screen angle with a limited choice. Transfer non-uniformity may cause bad graininess and an unsaturated color image. Geometric settings of the photoreceptor and transfer device and differentiation of electrostatic force applied to each color are effective to improve it. Hollow character is one of the defects in roller transfer. Toner image is compressed by mechanical pressure between a photoreceptor and a transfer roll in the transfer nip area. After the nip, a part of the toner is taken off by the cohesive force between toner particles and adhesive force with the photoreceptor surface (Figure 5.10 and Figure 5.11).[10]

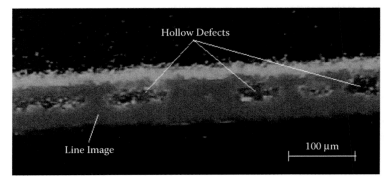

FIGURE 5.10 Hollow character.

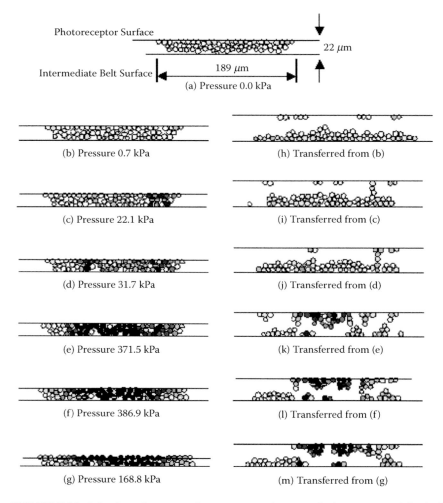

FIGURE 5.11 Adhesion phenomena between toner image and photoreceptor (physical simulation).

Effective methods to avoid this problem include driving the photoreceptor and the transfer roll at different surface speeds and using some additive on the toner surface to improve its fluidity. The intermediate transfer method is good at handling various papers in color printers, because of its weight range of 60–220 g/m² and because it reduces transfer action to the media to one time. The intermediate transfer method is becoming mainstream in color printers. The material for the intermediate transfer drum is elastic rubber, and the material for the belt type is a thin plastic film that is flexible enough to track photoreceptor shape. Both types have an electric resistance range of $10^8 \sim 10^{10}$ Ω cm semi-conductive to reserve appropriate transfer voltages (Figure 5.12).[12]

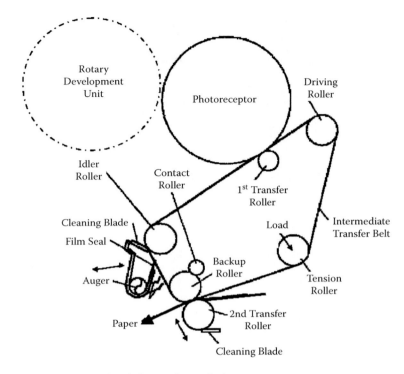

FIGURE 5.12 Intermediate belt transfer method.

Recent technology progress in this transfer process is remarkable for digital color image reproduction. However, there are still many issues to resolve, for example, its mechanical simplicity, stability, etc.

5.2.3.6 Fusing

The fusing process is to melt and fix a toner layer on a medium. Pairs of roll fusers are normally used. Rolls, such as fuser rolls and pressure rolls, contact each other to form a nip. Toner image on a medium passes through the nip. Thermal energy and mechanical pressure are impressed on the toner image in the nip. After the nip it is cooled to room temperature and the toner image is fixed. Therefore, an electronograph consists mainly of thermoplastic materials.

To achieve a maximum color gamut with a given toner set, the fused toner surface should be as smooth as possible. A smooth and soft surface fuser roll is suitable for this purpose.

On the other hand, this type of fuser roll fades quickly. Also, the peeling force is higher than on a rough surface, so toner offset from the media to the fuser roll. To achieve both longer life and a lower peeling force, more fuser oil should be applied or more wax should be added to the toner. The former case produces many side effects, so the latter case is better.

Microscopic random agglomeration of adjacent toner in the fusing process makes the graininess level worse. Agglomeration of the toner occurs if the viscosity of the toner at the fusing temperature is too low, so the adjacent toners literally stick to each other.

The binder resin's glass transition temperature T_g should be high enough for storage and the melting point T_m should be as low as possible for conservation of energy. Normally, T_g is about 50–70°C. Polyester and polystyrene are popular for binder resin. Viscosity and elasticity are both important for avoiding toner offset on the fuser roll. The dynamic shear elastic modulus range should be 2×10^6 to 5×10^5 dyn/cm^2 under a shear velocity of $10^2 s^{-1}$ corresponding to the effective fusing time.[13] Usually, toner material is designed to have two peaks of molecular weight distribution; the lower peak is for bonding between the toner and the medium, and the higher peak is for avoidance of hot offset of the toner layer. Polypropylene wax is sometimes used as a release agent.

Conservation of energy is now a big challenge for fusing; 60–80% of the total fusing energy consumption is spent just for standby. Thus, challenging points include a lower standby temperature and a faster revival time. Double roll-fusing and belt-fusing system geometries are shown in Figure 5.13, and a comparison of fusing technology in terms of energy consumption is shown in Table 5.3.[14]

5.2.3.7 Cleaning

After the transfer process, the cleaning device cleans the remaining toner on the photoreceptor surface for the successive electrophotographic cycle. Blades and brushes are popular, and a blade is usually used. The blade material is polyurethane resin, which is mechanically and chemically (ozone) strong. The cleaning process has great effect on image quality, such as background image density and image defects. However, the toner particle becomes smaller in size and spherical

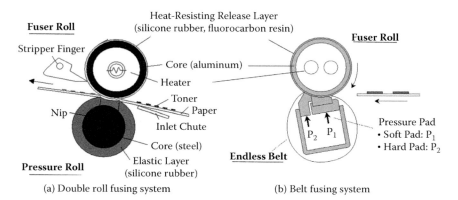

(a) Double roll fusing system   (b) Belt fusing system

FIGURE 5.13 Typical fusing systems.

TABLE 5.3
Comparison of Fusing Technology in Terms of Energy Consumption

Type	Power Consumption	Warm Up Time	Processing Speed	Thermal Efficiency	Disadvantage
Heat roller	Middle	Long	High	High	Neither good nor bad
Heat belt	Middle	Short	Middle	High	Belt drive stability temperature
Radiant	High	Short	Low	Low	Lowest thermal efficiency
Flash	High	Short	High	Low	Big and heavy device High power consumption
Pressure	Low	Zero	High	No need	Low fixing quality Heavy device weight High torque to drive

form to get higher print quality, and the smaller and smaller, and cleaning becomes more and more difficult in terms of photoreceptor life and overload, resulting in a banding defect.

5.2.3.8 Process Control

The electrophotographic process needs a certain precision of process control devices to stabilize its temperature/humidity dependency and cyclic instability. Electrostatic charge on the photoreceptor is discharged according to its resistance, and the resistance is a function of temperature. Powder toner holds an electrical charge and is discharged due to the surrounding humidity. So, temperature and humidity variation strongly affects the electrophotographic process. Control targets include laser ROS exposure intensity, electrostatic voltage on the photoreceptor, developed/transfer toner amount, toner concentration in development housing, and fused image density. The required precision depends on the application. One stressful application, for example, is graphic arts, because it requires higher color stability and faster speed, which consume toner rapidly. Figure 5.14 shows an example of a process control system.[15]

Optical density data of developed toner on the photoreceptor and a pixel count of original images feed back to the toner dispense motor to add an appropriate amount of toner supplied into the development housing. The optical density information is used to control applied voltages of electrostatic charges or to vary exposure intensity of laser ROS.

5.2.4 IMAGE QUALITY

Due to the increase in electronic originals, the relative amounts of printer output of all hard copies have been increasing. Since images created by PCs are noiseless, defects or non-uniformity standards required for printers are higher than those for copiers.

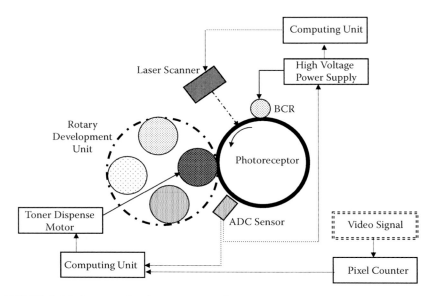

FIGURE 5.14 An example of process control systems.

With technological progress of high-definition displays and ecological trends, temporary information is sometimes processed only on displays, with no hard copies being made. For these reasons, higher image qualities are demanded for printers. Needless to say, preventive measures for negotiable securities, counterfeits, and copyright infringement will become vital.

To combine various imaging devices for creation and transmission of high-quality images, a method for reducing machine-to-machine color difference, day-to-day color instability, and also image deterioration at the time of compression/expansion will be required. Also, the color reproduction range needs to be expanded, and image/color management with enhanced accuracy is required. Color management includes gamut mapping, the color appearance model, measures on flare influence to display, and profile connection space (due to the differences in paper, UCR method, and colorant).

5.2.4.1 Color Fidelity/Color Stability

For color management in the office, it is desirable to achieve color fidelity regardless of which device to use (without designating the specific device). Required color fidelity levels vary according to image types. Examples show that the color tolerance range of skin color for natural images is CIELAB $\Delta E \leq 2$ for the green–red direction and CIELAB $\Delta E \leq 3$ for the blue–yellow direction.[16] Because characters have been output mostly in black and white, density stability in a high-density area has been critical. On the other hand, because pictures and diagrams, rather than characters, are output in color, it is critical to ensure density

TABLE 5.4
Comparison of Color Stability in Various Marking
Technologies (Day-to-Day Color Stability: CIELAB ΔE)

Lithographic Printing	Electro Photography	Sublimation Thermal Transfer	Inkjet	Thermal Transfer
3–5	5	1–2*	1**	2*

Values are measured on mid-level density gray path.

* CIELAB ΔE of color variation caused by donor roll replacement is about 4–5.
** Value when ink tank and marking head are not replaced.

stability ranging from low-density areas to high-density areas. Table 5.4 shows the comparison of color stability with various marking technologies.[17]

Table 5.4 indicates that currently, electrophotography needs measures to reduce day-to-day color instability. Table 5.5 shows examples of variation factors and amounts of each subsystem at a low-density area in electrophotography.[18]

The illuminated light intensity variation of the image input terminal (IIT) and the screen generator (SG) variation of the image output terminal (IOT) have a high contribution ratio to density variation at low-density areas. Table 5.6 shows a comparison of analog black-and-white, digital black-and-white, and digital color copiers.[18]

Due to digitization, the number of variable elements increases. For example, ROS and SG are the new variables. For color output, three signals (red, green, blue (RGB)) are necessary and three to four colors are required for developer units. For color, because the variation tolerance range will be twice as narrow, the difficulty of achieving the image quality target is estimated to be more difficult.

5.2.4.2 Color Uniformity

In electrophotography, color uniformity as well as color stability needs further improvements. As shown in Table 5.5, non-uniformity of illuminated light intensity, drum to wire space (DWS), electronic potential of photoreceptor, drum to roll space (DRS), and mass on sleeve (MOS) (amount of developer on developer roll) has a high contribution ratio to density non-uniformity for one copy. Measures to prevent cost increases while increasing accuracy need to be taken.

5.2.4.3 Tone

To reproduce high image quality, tones should be reproduced from 10% area coverage in the highlighted area, and the maximum density should be ≥ 1.8. Because defects such as rosette and moiré occur when reducing or enlarging

TABLE 5.5
Variation Factors and Amounts of Each Subsytem at Low-Density Area (10% Area Coverage)

Subsystem	Parameter	Density Nonuniformity	Density Variation
IIT	Illum. light intensity	± 0.013	± 0.021
	White reference plate	± 0.010	—
IOT SG[1]	Drift	—	± 0.020
IOT ROS	Light intensity	−0.003	± 0.001
IOT (charge)	DWS[2]	± 0.005	—
	GDS[3]	−0.002	—
IOT (P/R)	Background potential	± 0.008	± 0.008
	Dark potential	± 0.006	± 0.006
IOT (develop)	Toner density	± 0.001	± 0.002
	DRS[4]	± 0.005	—
	MOS[5]	± 0.005	—
IOT (transfer)	Transfer efficiency	± 0.002	± 0.001
System total		± 0.024	± 0.055
Target level		≤ 0.016	≤ 0.028

[1] SG = Halftone screen generator
[2] DWS = Drum to wire space
[3] GDS = Grid to drum space
[4] DRS = Drum to roll space
[5] MOS = Mass on sleeve

TABLE 5.6
Comparison of Analog Black-and-White, Digital Black-and-White, and Digital Color Copiers

	Analog Black-and-White	Digital Black-and-White	Digital Color
Input	Illumination optical system	Illumination optical system	Illumination optical system RGB separation
Output	XERO	ROS SG XERO	ROS SG XERO DEVE 3–4
Number of variable element	2	4	6–7
Required accuracy	Same	Same	Approximately twice
Difficulty	Same	Twice	Approximately six times

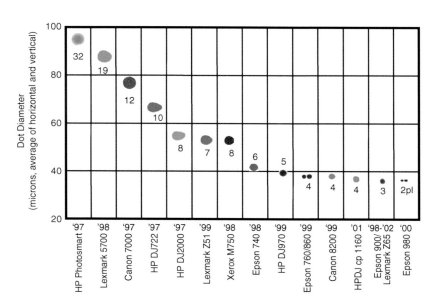

FIGURE 4.2 Evolution of drop volume and drop size.

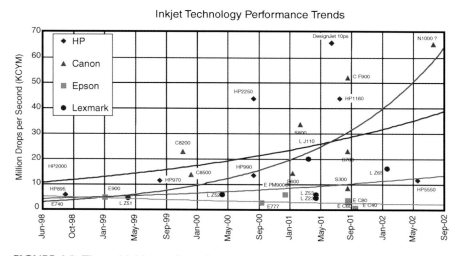

FIGURE 4.3 Thermal inkjet configuration.

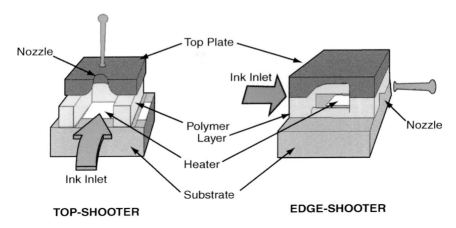

TOP-SHOOTER   **EDGE-SHOOTER**

FIGURE 4.4 Thermal inkjet drop ejection process.

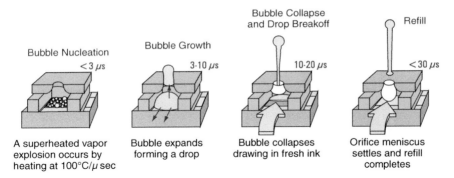

Bubble Nucleation
< 3 μs

Bubble Growth
3-10 μs

Bubble Collapse and Drop Breakoff
10-20 μs

Refill
< 30 μs

A superheated vapor explosion occurs by heating at 100°C/μ sec

Bubble expands forming a drop

Bubble collapses drawing in fresh ink

Orifice meniscus settles and refill completes

FIGURE 4.5 Top view of particle-tolerant ink supply channels.

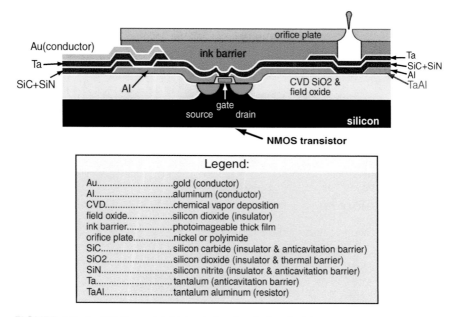

FIGURE 4.7 An HP thermal inkjet printhead and electrical interconnect.

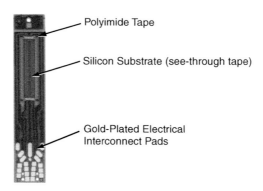

FIGURE 4.8 Epson MLP piezo inkjet drop generator.

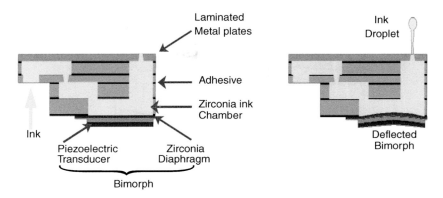

FIGURE 4.10 Drive waveforms for variable drop volumes.

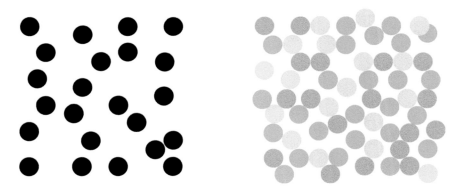

FIGURE 4.13 Image processing pipeline.

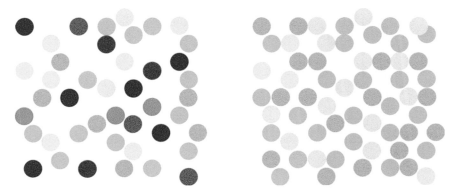

FIGURE 4.16 Gamut of CMY and CMYGB printers.

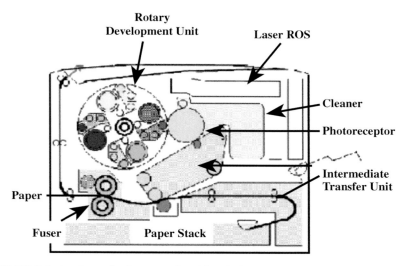

FIGURE 5.4 A schematic diagram of multi-path intermediate belt transfer color laser printer.

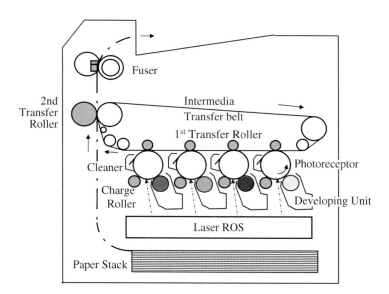

FIGURE 5.5 A schematic diagram of tandem intermediate belt transfer color laser printer.

FIGURE 7.7 Film scanner unit of a Frontier 330 (area CCD).

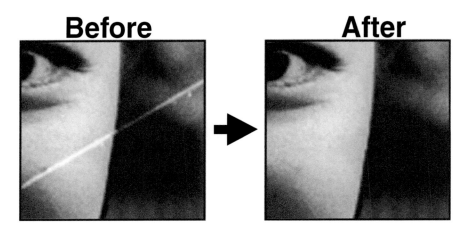

FIGURE 7.10 Automatic scratch and dust restoration function.

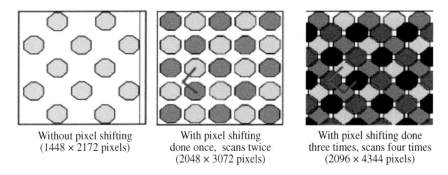

Without pixel shifting
(1448 × 2172 pixels)

With pixel shifting
done once, scans twice
(2048 × 3072 pixels)

With pixel shifting done
three times, scans four times
(2096 × 4344 pixels)

FIGURE 7.11 Pixel-shifting method.

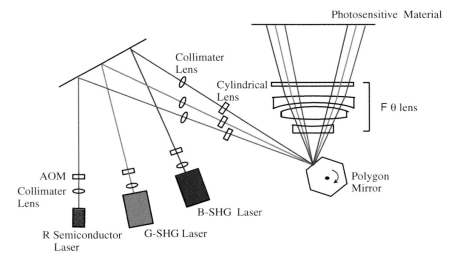

FIGURE 7.13 Optical system for laser exposure.

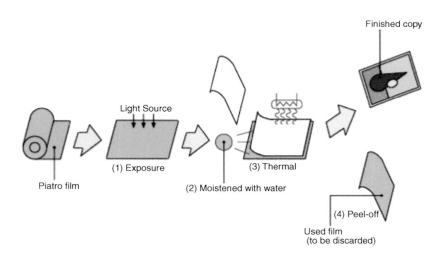

FIGURE 7.14 Conceptual diagram of pictrography.

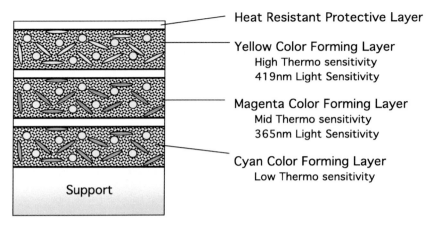

FIGURE 7.18 Basic structure of TA paper.

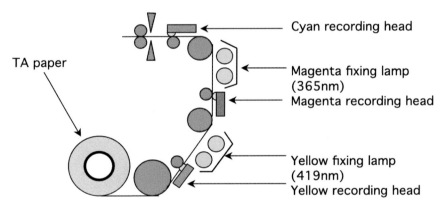

FIGURE 7.25 Basic configuration of a high-speed, three-head tandem digital color printer.

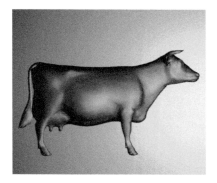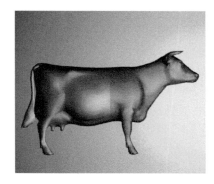

FIGURE 9.2 Garrett Johnson's "Metameric Cows" [Johnson, 1998]. This demonstration simulates how the same reflectance properties under a single light source can appear different to different observers. The front half and the back half of the cow are different reflectances. To the two-degree observer (left), the front and the back have the same color. To the ten-degree observer (right), the front and the back have different colors.

FIGURE 9.5 Spectral color management would provide a way to impose a new light source on the image of a captured object.

images, it will be critical to increase the number of lines. Inkjet manages to improve both tone and resolution using various measures.

In current electrophotography, the typical halftone screen frequency is 200 lines/inch and the typical depth for each halftone screen cell is 8 bits. It should be noted that if one wishes to achieve higher resolution image reproduction, it requires not only adoption of a higher halftone screen frequency, but also improvement in microscopic image reproduction accuracy. Otherwise, it creates more microscopic image noise and a higher gammer tone curve.

5.2.4.4 Reproduction of Fonts

In lithographic printing, different fonts are used for different purposes. 1800-dots-per-inch (dpi) lithographic printers for the office, with 14 types of Japanese fonts and 64 types of European fonts, are already available. The number of users who select various types of fonts has gradually been increasing. Some use unique fonts to express originality. However, it is challenging to reproduce, in one page, high image quality text/image mixed originals in various fonts sent from various recording media or digital cameras. It is also important to offer products of this capability in prices available for general offices. Thus, more improvements are vital for image formatting and speed-up of image processing. Marking technology varies from font to font. In digital lithographic printing, marking technology ranges from 1200 dpi to 4000 dpi (though they may vary according to text size). This is considerably higher than current non-impact printers. Table 5.7 shows difference of aspect ratios among various fonts.

This example shows that in order to reproduce the aspect ratio accurately, it is necessary to adjust the line width by approximately 22 μm for 8-point characters. Font form and density reproduction must be identical when printers of different manufacturers or types are used to print electronic originals. Thus, technology to adjust difference in output among printers has come to be the critical issue.

TABLE 5.7
Difference of Various Fonts' Aspect Ratios

Font	Aspect Ratio	Aspect Difference of 8-Point Character
Socho	1:0.71	0.044 mm
Kyokasho	1:0.60	0.088 mm
Kaisho	1:0.50	0.110 mm
Fantail	1:2.00	0.110 mm
Gothic	1:1.33	0.044 mm

5.2.4.5 Defects

In network systems, it is also necessary to consider elements other than marking technology. Image deterioration (a defect which does not exist in the original image) caused by image compression or expansion (e.g., J-PEG) becomes a critical issue when retrieving data stored in memory and then copying (e.g., electronic pre-collate copy (EPC)) or printing via network. Mosquito noise, such as random small mosquito shape artifacts, is one example of this type of image deterioration.

5.2.4.6 Graininess

Graininess characteristics vary for each marking technology. Lowest graininess can be achieved with the sublimation thermal transfer method. Because it is easy for inkjet to maintain the same size of ink particle, inkjet has an advantage compared with other marking technologies. With thermal transfer printing (TTP), blotch occurs in the secondary color caused by the boundary of primary and secondary colors. Because graininess is high for two-color and three-color, it is vital to make the ink donor film thinner. In electrophotography, graininess is improved with smaller toner particles.[19–21] Figure 5.15 shows an example.[22]

Figure 5.15 indicates that low graininess can be obtained if the toner particle diameter is smaller than 2 μm, regardless of ROS beam diameter or number of lines. In order to obtain an ideal particle property when the toner particle diameter is larger than 2 μm, it is desirable to keep the ROS beam diameter small and minimize the number of lines within the range so image structure cannot be recognized as a visual characteristic.

5.2.4.7 Banding, Colors to Colors Miss-Registration

Stripes of density non-uniformity that occur in the uniform shading area in tables or graphs or white gaps between text and shading background caused by the colors to colors miss-registration (Figure 5.16) greatly influence the quality of business documents.

The striped patterns called *banding* are normally observed at the 0.2–2 cycle/mm spatial frequency where the human visual system has high sensitivity. Figure 5.17 shows that banding is visually recognized most easily at about 1 cycle/mm spatial frequency.[23] A report shows banding is also dependent on the number of lines. With 200 lines 8 bits, 0.5% peak-to-peak frequency is necessary for a spatial frequency of 0.38 cycle/mm.[24] In the printing industry, 25–30 μm[25] colors to colors misregistration is necessary in a high-quality image.

With inkjet, not only colors to colors miss-registration but also stitching noise, which occurs in the process direction with the interval of the ink head, have been the problems to be solved. However, by changing the marking order, they have improved dramatically.

In digital electrophotographic technology, the main factors that cause banding or colors to colors misregistration are positional inaccuracy when writing in light

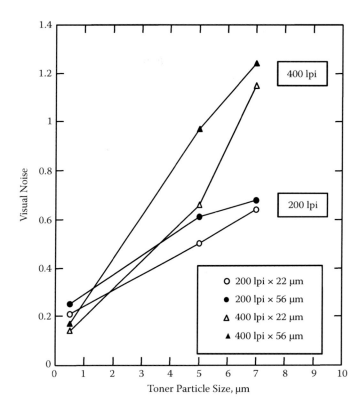

FIGURE 5.15 Effect of toner particle size, screen frequency, laser ROS beam diameter in graininess.

signals to photoreceptors by ROS and zigzag motions or velocity variation of photoreceptors/transfer belt.

Factors for fast scan direction positional inaccuracy when writing in light signals by laser ROS are considered to be rotational inconsistency of polygonal mirror or jitter.[26] The photoreceptor velocity varies according to the photoreceptor's gear pitch in the mechanical driving system, the number of teeth, or the source resultant pulse number for motor excitation. How to prevent transition of unique vibrations for the driving transfer system configuration factor and how to reduce vibration transfer sensitivity were presented in a previous study.[27]

5.2.4.8 Color Gamut

Figure 5.18 shows the color gamut of color copiers that are already commercialized. To incorporate a large color gamut of red and blue areas, it is effective to make the bevel of magenta colorant on the long wavelength sharp and the reflectance maximum on the short wavelength.[28] Such measures are now being discussed (Figure 5.19).

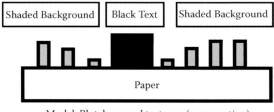

Model: Blotch around text area (cross section)

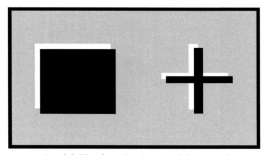

Model: Blotch on border area (plain view)

FIGURE 5.16 Examples of defects in print.

5.3 TONER

5.3.1 TYPES OF DEVELOPER (TONER AND CARRIER)

Developer for laser printers is categorized in Figure 5.20. Powder and liquid are the primary classifications. Powder developer can be categorized into two types, single-component and two-component, by the difference in the triboelectric charging method.

In single-component developers, the toner obtains its triboelectric charge by interacting with the development sleeve surface and the trimming blade (see Figure 5.9b).

In the case of magnetic single-component developer, the toner particles are attracted onto the development sleeve by magnetic force. In the case of a non-magnetic developer, the toner particles are transported to the development zone by an electrostatic force.

Two-component developers consist of carrier beads and toner. These particles are mixed together and generate a triboelectric charge. Usually, magnetic powder is used as a carrier. The electric property of the carrier beads affects development characteristics. Development with insulating carrier beads is good for fine-line image reproduction, and development with conductive carrier is good for large-area solid image reproduction. The difference is because of the strength of the

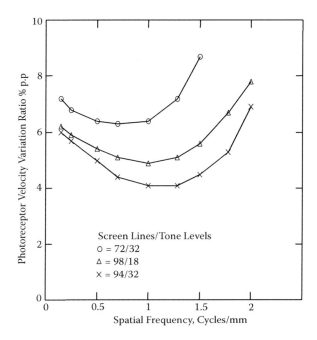

FIGURE 5.17 Acceptability of banding level.

effective electric field in the development area and the residual electric charge in carrier beads after development.

Liquid developer consists of toner dispersed in insulating liquid (carrier). Insulating carrier is organic solvent, and toner particle size is smaller, i.e., 1–2 microns, than powder developer, which gives better image quality. Thus, liquid developer is suitable for publishing laser printers.

Characteristics of developers are shown in Table 5.8.

Developer should satisfy various functions such as storage stability, environmental protection (safely disposable and recycling use), and electrophotographic capabilities. Environmental impact and influence become more important year by year in toner material design.

5.3.2 INGREDIENTS OF TONER AND CARRIER

Toner is for visualization of images, so colorant and binder resin are obligate ingredients. Carrier is for tribo electric charging and retention, and held and feed by magnetic force, so magnetic material and surface coating resin to control tribo electric charging are obligate ingredients.

Figure 5.21 shows the toner surface by electron micrograph.[29] A powder toner particle is 5–12 microns in diameter, and the shape is a round or indeterminate form, depending on its manufacturing method. Many toners consist of binder

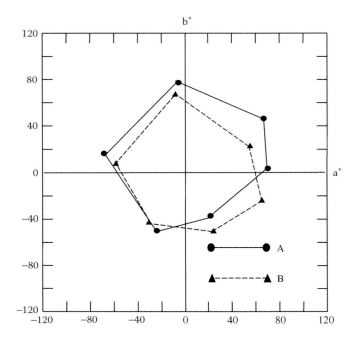

FIGURE 5.18 Color gamuts in color copiers.

resin, colorant, charge control agent, and release agent. In many cases, the surface has an external additive for its fluidity and charge control.

Figure 5.22 shows a photograph of a ferrite carrier core whose surface is coated by resin to control electric resistance.[30]

5.3.2.1 Toner

5.3.2.1.1 Binder Resin

Binder resin bonds toner onto media to form visible images. This bonding and preventing from fading are main functions of the binder resin. However, binder resin is a fundamental material, because its characteristics affect every electro-photographic subsystem.

Resin is a polymerized monomer, and composition of monomers can control resin characteristics. Molecular weight distribution of resin is controlled to have an appropriate bonding and releasing characteristic.

Figure 5.23 shows a toner viscoelastic behavior conceptual diagram. Fusing of toner image is done in the elastomeric region. If the toner temperature is lower than the elastomeric region, the toner is not able to fix on a medium. If the temperature is in the fluid region, it is easy to cause a hot offset problem (toner image sticks to a fusing roll). Therefore, keeping the elastomeric region as wide as possible is necessary to prevent unfusing and hot offset problems.

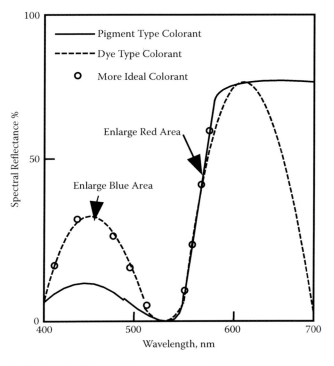

FIGURE 5.19 Schematic diagram of magenta spectral reflectance curve.

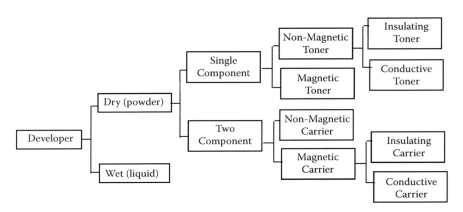

FIGURE 5.20 Classification of developer for laser printers.

Resin with multiple peaks in molecular weight distribution is a mixture of a high-molecular-weight component and a low-molecular-weight component. The low-molecular-weight component improves bonding characteristics, and the high-molecular-weight component prevents toner offset.

TABLE 5.8
Characteristics of Developers

Developer	Advantage	Disadvantage
Two-component	Easy to control tribo electric charge High-speed develop ability Easy-to-use color toner Less environment dependent	Need to control toner concentration Complicated structure of deve. Unit developer needs to be exchanged.
Magnetic single-component	Simple structure of development unit Free to control toner concentration	High accuracy is required Difficult to use color toner Difficult to control tribo electric charge Environment dependent
Non-magnetic single-component	Simple structure of development unit free to control toner concentration Easy-to-use color toner	Difficult to control tribo electric charge Difficult to improve speed of development

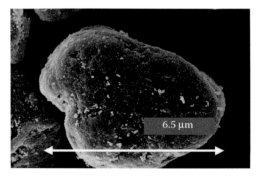

FIGURE 5.21 Toner surface view.

Lowering fusing energy can be done by lowering molecular weight. However, lowering molecular weight lowers the resin glass transition temperature and causes bad storage stability. Therefore, not only lowering molecular weight but also changing the composition of resin has been implemented.

Typical resins for the heat-conduction fusing method are styrene–acrylic copolymer, styrene–butadiene copolymer, and polyester.

Styrene–acrylic copolymer resin is inexpensive because of its monomer composition, which has triboelectric charging control ability and ease of molecular weight control.

Polyester resin has high mechanical strength and superb viscoelasticity characteristics, but it is more expensive than styrene–acrylic copolymer resin in general.

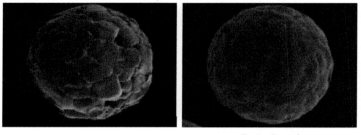

Resin Coated

FIGURE 5.22 Ferrite carrier.

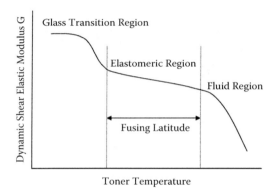

FIGURE 5.23 Toner viscoelastic behavior.

Styrene–butadiene copolymer resin is also inexpensive; however, the molten viscosity of it is relatively low, so silicone oil and its applicator mechanism are obligate.

5.3.2.1.2 Other Typical Construction Materials of Toner

Toner consists of various materials besides binder resin to achieve many electrographic functions. Colorant, charge control agent, release agent, and external additive are typical secondary toner materials. The main compositions are shown in Table 5.9.

5.3.2.2 Carrier

The basic functions of the carrier are to charge toner particles with the triboelectric charging process and to hold toner particles on its surface until it reaches the development zone.

The magnetic developer feeding mechanism and magnetic carrier are commonly used. The magnetic carrier is categorized into two types: magnetic material

TABLE 5.9
Typical Construction Materials of Toner

Item	Function	Classification	Typical Materials	Content	Remarks
Colorant	Black	Carbon black (non-magnetic toner)	Furnase black, channel black	5–10 wt%	Carbon black is sometimes used for magnetic toner as an electric resistance adjuster.
		Magnetic particle (magnetic toner)	Magnetite (100–500 nm) cubic, spherical, polyhedral structure	20–80 wt%	Spherical shape looks red-tinged black and cubic shape has a bluish tinge.
	Color	Yellow	C.I. Pigment Yellow 17 Non-benzene yellow pigments	5–10 wt%	To achieve a good transparency, colorant should be dispersed evenly and finely in toner resin. Surface treatment or processed pigment to meet dispersion criteria.
		Magenta	Quinacridone carmine 6B		
		Cyan	Copper–phthalocyanine		
Charge control agent	Plus charging	Nitrogen compound	Nigrosin (for black toner) 4 class ammonium salt series (color toner)	1–5 wt%	4 class ammonium salt series charge control agents have no color and are transparent.
	Minus charging	Halogenated compound and metal complex	Azo series or salicylic series metallic complex		Non-metallic complexes are approached for heavy metal free movement.
Release agent	Lowering critical surface tension of toner	Olefin series wax Ester series natural product	Polypropylene and polyethylene Carnuba wax	0–5 wt%	Low critical surface tension, melting at low temperature sharply and high hardness under room temperature.
External additive	For fluidity improvement For cleanability	Fine inorganic particle Higher fatty acid metallic salt	Silica (about 7–15 nm) Zinc stearate	0–5 wt%	Other function of external additive is to improve tribo electric charging characteristics of toner.

coated with resin and magnetic material dispersed in resin. Magnetic materials for the carrier are iron, magnetite, copper–zinc ferrite, manganese ferrite, and light metal ferrite.

Resin coating provides an electrostatic property and electric resistance adjustment. It prevents toner contamination onto the carrier surface as well. Resin coating materials include styrene–acrylic copolymer, fluorine series such as vinylidene fluoride, polyethylene, and silicon resin.

The magnetic material dispersed resin type carrier is manufactured like magnetic toner; the resin and magnetic material melt together, knead well, cool down, and then smash up. The smaller particle size and lighter weight are selling points for this type of carrier.

5.3.3 TONER MANUFACTURING PROCESS

The schematic flow of a conventional pulverizing process is shown in Figure 5.24.[31] First, the toner components of resin, colorant, and other materials are mixed in a dry process (pre-mixture process). Subsequently, the resin is softened by applying heat and kneading it with strong shearing stress, and the resin bulk or pellet is obtained that has a uniform distribution of the component, such as colorant (knead process). Then it is pulverized to several micron meter-sized particles (pulverization process). The air jet pulverizing method uses high-speed airflow, and the mechanical pulverizing method uses shear stress. Finally, toner is obtained through a classification process that takes out the desired particle diameter from the obtained pulverized material (classification process). The rough

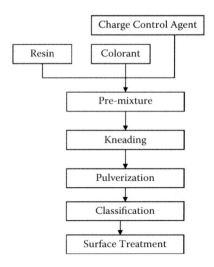

FIGURE 5.24 Process flow of pulverized toner.

powder and fine powder that are generated in this process are collected and reused as raw material. After that, some external additives are mixed and put into a container.

Polymerization is a method of manufacturing, including the stage of synthesizing resin that is different from the conventional pulverizing process. This process can be roughly classified into suspension polymerization and emulsion aggregation. In the suspension polymerization process, the contents of the toner component such as colorant are dispersed in a monomer. The droplet of a predetermined size as a toner is made from the monomer composite that is suspension dispersed in water using a dispersion stabilization agent.

Subsequently, the coloring particle that dispersed the content of the toner composition ingredient is formed by suspension polymerization. After removing the dispersion stabilization agent, filtration, washing, drying, and mixing with external additives, toner for the laser printer is obtained.

The schematic flow of this manufacturing process is shown in Figure 5.25.[32] Because a droplet is formed in water by the suspension polymerization method, the droplet (toner) becomes a spherical shape.

Considering the principle of electrophotography, the spherical shape toner has an ideal performance. However, it has problems such as poor cleaning of the remaining toner because of the larger adhesion force to a photoreceptor.

The emulsion aggregation manufacture process is also put into practical use as a method to form a semi-spherical shape toner by the polymerization method. Figure 5.26 shows the schematic flow of toner preparation.[33]

Emulsion aggregation toner may be prepared by using the following steps. First, there is pigment/wax dispersion preparation (dispersed in de-ionized water), and emulsion polymerization (obtained latex is about 200 nm in size). Latex size can be controlled by surfactant concentration, homogenization (the mixture is homogenized with a high shear mixer; average size is about 2.5 microns after

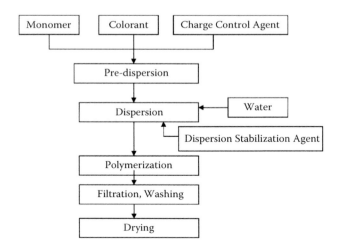

FIGURE 5.25 Process flow of suspension polymerization toner.

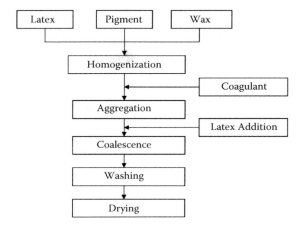

FIGURE 5.26 Process flow of emulsion aggregation toner.

homogenization), aggregation (the homogenized mixture is heated with continuous mixing in the reactor and with temperature ramping), and coalescence (the pH of the slurry is controlled to freeze the aggregated size and then the mixture is ramped over the glass transition temperature and held for several hours with mixing). After cool down, the filtered coalesced particle is washed with de-ionized water and dried. Narrow size distribution can be obtained by optimization of the latex size and Zeta potential, controlled in the emulsion polymerization step (Figure 5.27).[34]

High color image quality may be achieved by fine image reproduction and wide color gamut based on small particle size, narrow distribution, and the optimization of fusing properties with low-molecular-weight polyethylene wax incorporation.

Although it is not a polymerization method, another toner-manufacturing process, without passing through pulverization process, has also been proposed. Resin, colorants, and other materials are dissolved and dispersed in solvent that is insoluble in water. The solvent is dispersed in water to form oil drops of toner size. Removing a solvent forms a toner particle. In this case, the resin can use various kinds (i.e., polyester), because this process can use materials that polymerized beforehand. Figure 5.28 shows three kinds of toner surfaces manufactured by various processes

5.4 MEDIA AND CONSUMABLES

Media, toners/developers (see Section 5.3), and photoreceptors (see Section 5.2.3.1) are the major consumables for color laser printers. Laser color printers should be able to handle any kind of medium. However, there are several aspects that are suitable to color laser printers. These include electrostatic stability with temperature and humidity, appropriate elasticity for feeding, and resistance to

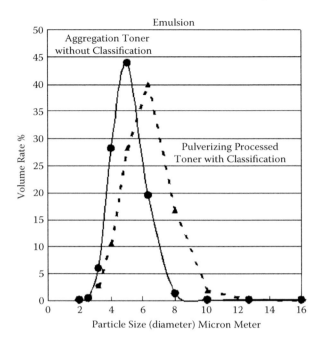

FIGURE 5.27 Particle size distribution of emulsion aggregation toner and conventional toner.

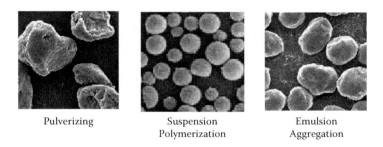

FIGURE 5.28 Comparison of toner surface.

curl with toner and media thermal contraction after fusing. Unique aspects for transparency media are false contour and transparency projection quality. If the transparency media surface is not soft enough to penetrate the toner image, the edge of the toner image will turn out to be a bump, reflecting incident light and becoming a dark band looking like a false contour. Transparency media projection images sometimes look dull, dark, and less saturated. This occurs when either the toner is not well melted in the fusing process or the transparency medium itself has low transparency.

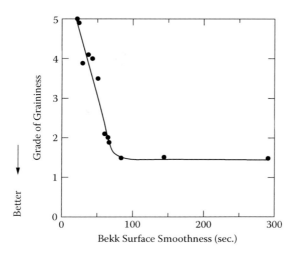

FIGURE 5.29 The relationship between paper surface smoothness and halftone graininess.

An image quality relating aspect is paper surface roughness, as shown in Figure 5.29, where smoother paper gives less graininess, which is an important attribute for color.[35] However, smoother paper has high material density that is too soft for feeding and too thin (with the same paper weight) to obtain enough opacity for showthrough. Thus, for colored paper, heavier weight with more filler, such as potassium carbonate added to it, is preferred. Smoother paper is also easy to curl. Additional tweaking such as optimizing paper fiber alignment, reduces paper curl. Figure 5.30 shows the relationship of the paper fiber orientation ratio

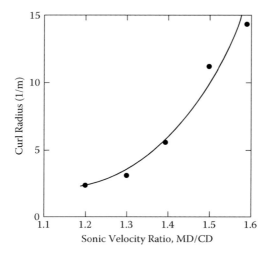

FIGURE 5.30 The relationship between fiber orientation ration and radius of post-fuser-curl.

(MD [machine direction]/CD [cross direction] of a paper pressing machine and paper curl under stress conditions (high image area coverage).[36]

The customer replaceable unit (CRU) and service engineer replaceable unit (ERU) are other consumables for color laser printers. CRU and ERU are cartridges usually consisting of a toner bottle, photoreceptor, development housing, and cleaner that give better maintainability.

REFERENCES

1. JEITA research report, JEITA, 04-P-2 (2004).
2. Dessauer, Clark, *Xerography and related processes*, The Focal Press, Burlington, MA (1968).
3. Schaffert, R.M., *Electrophotography*, The Focal Press, Burlington, MA (1975).
4. Fukase, Yasuji, *Journal of Printing Science and Technology*, 33, 2, pp. 29–35 (1996).
5. Kimura, Kiyoshi et al., Color Laser Wind 3310, Fuji XEROX Technical Report No. 12, p. 161 (1998).
6. Tamura, Kazuo et al., DocuPrint C2220/C2221, Fuji XEROX Technical Report No. 14, pp. 104–115 (2002).
7. Uchida, Teiji, Optics for engineers, *The Journal of the Institute of Telecommunications Engineers*, 62, 5, pp. 538–545 (1979).
8. Kawamura, Naoto, Digital marking technology, *Journal of The Society of Electrophotography of Japan*, 26, 1, pp. 75–82 (1987).
9. Kataoka, Keizo, High speed and high resolution laser scanning optics using multi beam laser, *Journal of The Imaging Society of Japan*, 37, 1, pp. 91–98 (1998).
10. Fukase, Yasuji et al., Digital Hardcopy Technology, Kyoritsu-Shuppan Co., Ltd., p. 199 (2000).
11. Nakayama, Nobuyuki et al., Numerical Simulation of Electrostatic Transfer Process Using Discrete Element Method, PPIC/Japan Hardcopy Conference '98 Proceedings, pp. 261–264 (1998).
12. Okamoto, Yoshikazu et al., Intermediate Transfer Belt System for Color Xerography, Fuji XEROX Technical Report No. 12, pp. 22–31 (1998).
13. The Imaging Society of Japan, Basics and Application of Electrophotography, Corona Publishing Co., Ltd., p. 197 (1988).
14. Matsumoto, Shinji, Energy Saved Fusing System, Technology Seminar of Fall Meeting, The Imaging Society of Japan (1999).
15. Kimura, Kiyoshi et al., Color Laser Wind 3310, Fuji XEROX Technical Report No. 12, p. 161 (1998).
16. Nakaya, Fumio, The 16th Fall Seminar of The Institute of Image Electronics Engineers of Japan Proceedings (1992).
17. Bolte, Steve, SPIE, 1670 Color Hard Copy and Graphic Arts, pp. 3 (1992).
18. Nakaya, Fumio, The 8th Conference of the Color Technology '91 Proceedings, pp. 141–148 (1991).
19. Shaw, R. P. Dooley, Noise perception in electrophotography, *Applied Photographic Engineering*, 5(4), pp. 190–196 (1979).
20. Yamazaki, T., et al., *The Journal of The Institute of Image Electronics Engineers of Japan*, 21, 2, pp. 106 (1992).

21. Chiba, T., The 4th NIP Symposium Proceedings, pp. 129 (1988).
22. Shigehiro, Kiyoshi, The Effects of Toner Particle Size and Image Structure on the Image Quality in Electrophotography, The 9th NIP Symposium Proceedings, pp. 97 (1993).
23. Ishikawa, Hiroshi et al., Effect of Photoreceptor Velocity Variation on Halftone Image Reproduction, The 3rd NIP Symposium Proceedings, pp. 133 (1986).
24. Hirakura, Koji, Drum Digital Color Electo Photographic System, Japan Hard Copy '91 Proceedings, pp. 101 (1991).
25. Kume, Masatsugu, Insatsu Joho, pp. 23 (1992).
26. Sakaue, Eiichi, Brief Summary and Outlook of Laser Direct Prepress, *The Journal of The Society of Electrophotography of Japan*, 33, 2, pp. 170 (1994).
27. Miwa, Tadashi et al., Konica Technical Report, 6, pp. 29 (1993).
28. Tsuda, Shinich et al., The 5th Information and Image Processing Conference of The Japan Society of Colour Material Proceedings, pp. 26–33 (1994).
29. Ichimura, Masanori et al., The Smallest Particle Size Full Color Toner, Fuji XEROX Technical Report No. 13, p. 170 (2000).
30. Yamazaki, Hiroshi, Basics of Developer Technology and Its Future Trend, 53rd Technology Seminar: The Imaging Society of Japan, p. 109 (2002).
31. The Imaging Society of Japan, Basics and Application of Electrophotography, Corona-Pub., pp. 482–485 (1988).
32. Yamazaki, Hiroshi, Basics of Developer Technology and Its Future Trend, 53rd Technology Seminar: The Imaging Society of Japan, pp. 118–120 (2002).
33. Matsumura, Yasuo et al., Encapsulated Emulsion Aggregation Toner for High Quality Color Printing, IS&T NIP17: International Conference on Digital Printing Technologies, pp. 341–344 (2001).
34. Matsumura, Yasuo et al., Development of EA Toner (Emulsion Aggregation Toner) for High Quality and Oil-Less Printing, Fuji XEROX Technical Report No. 14, p. 98 (2002).
35. Matsuda, Tsukasa, Fuji Xerox J Paper, Fuji XEROX Technical Report No. 7, p. 56 (1992).

6 Dye Thermal-Transfer Printer

Nobuhito Matsushiro

CONTENTS

6.1 INTRODUCTION

The principle of image formation by thermal-transfer printers is depicted in Figure 6.1. Figure 6.1 relates to the dye sublimation and the wax melt printer as most representative of thermal-transfer printers. The principle of thermal-transfer printers is that using a heating element, physical or chemical reactions of solid-state ink form images. An input voltage heats the thermal-generating elements. Onto the recording sheet, the base material sublimes or melts from the heated area of the ink.

In a thermal-transfer printer, an ink supply and ink recovery mechanism or clog recovery system are not required as in inkjet printers, and these are advantages of the thermal-transfer printer. The only mechanical parts required are the driving mechanism of ink sheets and coloring sheets, making the recording and printing mechanisms relatively simple. The product can be compact, lightweight, and inexpensive because the structure of the printer engine is so simple.

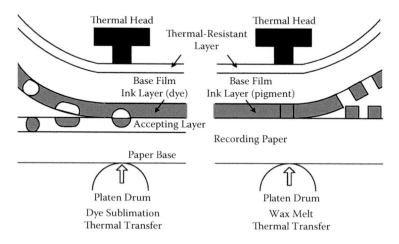

FIGURE 6.1 Dye thermal transfer and wax melt thermal transfer.

6.2 DRIVING MECHANISM

6.2.1 LINE SEQUENTIAL METHODS

The serial line sequential method and the parallel line sequential method are two driving methods (Figure 6.2). In the serial line sequential method, the recording sheet moves forward gradually, and the thermal head shifts for each color. In the parallel line sequential method, each color is aligned in parallel, the recording sheet is forwarded, and the thermal head is shifted in parallel.

6.2.2 AREA SEQUENTIAL METHOD

The processing speed by this method is high. Currently, most printers adopt this method. As shown in Figure 6.3, transfer sheets of the same size as the recording sheets are continuously aligned, and transferring is carried out onto the entire area of the sheet.

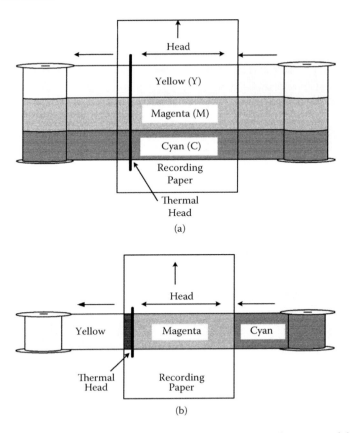

FIGURE 6.2 (a) Line sequential method (parallel sheet); (b) line sequential method (sequential sheet).

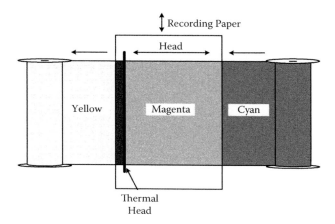

FIGURE 6.3 Area sequential method.

6.2.3 TYPICAL CONFIGURATION

A typical driving configuration of a thermal-transfer printer is depicted in Figure 6.4. The engine shown is the swing type. The recording paper returns to its initial position when one color is completely applied to the paper. Low expense and compactness are advantages of this type of printer. A drum-winding system and a three-head system also exist. In the drum-winding system, registration is relatively simple and the return of the recording sheet is not required; this contributes to a reduction in the recording time. The three-head system is suitable for high-speed recording.

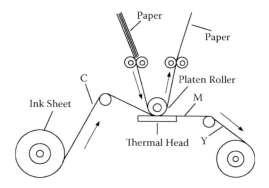

FIGURE 6.4 Typical configuration of a thermal-transfer printer.

6.3 DETAILS OF DYE SUBLIMATION PRINTER

6.3.1 BASIC STRUCTURE

The dye sublimination printer is one of the most promising printers of recent years. Sublimation is the cycle that forms the basis of dye sublimation printers. Sublimation is a process by which solids are transformed directly between solid and vapor without passing through a liquid phase. The outline structure of the dye sublimation printer engine is shown in Figure 6.1. Ink sheets are media onto which dye has been dispersed for subsequent transfer to the recording sheet.

6.3.2 SUBLIMATION DYE AND SHEETS

In a sublimation dye system, the energy required for transfer is greater than that used for a wax melt printer. The color carrier in the sublimation dyes must by design be unstable but, at the same time, recorded materials must be stable. This is contradictory to the nature of sublimable dyes. Accordingly, the selection of dyes is the most important factor in determining the success or failure of ink sheet manufacturing.

Reproduction of colors is realized by subtractive color mixture. Ink sheets used in the system have the three subtractive primary colors of cyan, magenta, and yellow, and often black is also used. To satisfy requirements such as light resistance and color reproducibility, special dyes were developed from the middle of the 1980s. At the beginning of the 1990s, new dyes with the same molecular structure as pigments used for color photosensitive materials were developed. Progress has been made in the area of materials development of dyes with appropriate subliming characteristics — characteristics for dispersing by diffusion, high absorption coefficient, severe weather tolerance, and good saturation.

The requirements for the dyes applied onto the recording sheet include high linearity with respect to the applied heat and many levels of gradation. The ink sheets should not degrade the thermal head, the ink sheets should not adhere to the thermal head, and no ink should remain on the thermal head. The thermal-resistant layer should not affect the properties of the dye layer. In the recording sheets, chemical compounds with high hardness are included.

It is essential that the ink sheets be made of thermal-resistant materials because the the the thermal head reaches 280–340°C momentarily. Only polyester films are available as low-cost sheets with thermal-resistant properties and strength. However, until recently, polyester films have had short useful lives under these conditions. In recent years, methods have been developed to incorporate a layer with improved longevity.

Recording sheets should not degrade the thermal head with the compounds when they are in contact. Recording sheets for dye sublimation printers are special media. There is no path currently foreseen to support plain paper.

6.4 DETAILS OF WAX MELT PRINTERS

6.4.1 BASIC STRUCTURE

The mechanisms of two thermal-transfer printers are compared in Figure 6.1. The major difference in terms of the thermal-transfer mechanisms of the two systems is that, compared with the wax melt printer, the dye sublimation printer requires higher thermal energy.

6.4.2 MELT WAX AND SHEETS

In the wax melt printer, a wax containing pigments is used as ink. Typical wax media such as paraffin are used. Coloring materials such as pigments are dispersed, and coated sheets of inks are produced. Pigments are used as the colorants. Pigments whose characteristics are very close to those of printing ink can be used. The wax functions as a carrier for the pigments during heating, melting, and transferring by the thermal head. When cooled, the dried wax fixes the pigments as a binder onto the recording sheets. As vehicles, paraffin wax, carnauba wax, and polyethylene wax are used.

The wax melt printer employs more stable dyes and pigments than does the dye sublimation printer for high-quality images. Many media, including plain paper, may be used for recording.

6.5 THERMAL HEAD

6.5.1 REQUIREMENTS

The thermal head must be durable throughout continuous rapid heating and cooling cycles. For high-speed printing, the thermal head must have thermal properties that allow rapid heating and cooling. Thus, the thermal head must have low heat retention capacity. To reduce power consumption, the heat energy supply must be highly efficient. For high-resolution printing, high-density resistor patterns must be realized.

6.5.2 STRUCTURES AND FEATURES

There are three types of typical thermal heads: thin-film, thick-film, and semiconductor. The thin-film head is suitable for high-speed printing because of its excellent thermal response. The structure of the head is shown in Figure 6.5(a). However, the production of large-sized thin-film heads is difficult, and the manufacturing process is complex. This head is frequently used in both sublimation and melt printers. The heating element is formed using screen-printing and sintering technology. The simple fabrication process is an advantage, and thus it is suitable for mass production.

A thick-film head is most suitable for large printer engines and used for large-sized sheet printers. In the case of sublimation transferring, the conduction time of the current to the thermal head is 2–20 ms, and that in melt transferring is

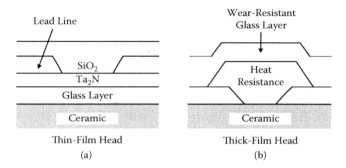

Thin-Film Head
(a)

Thick-Film Head
(b)

FIGURE 6.5 Thin-film head and thick-film head.

approximately 1 ms. The difference in duration arises from the difference in recording processes, reflected in the recording speed. Figure 6.5(b) shows the structure of thick-film heads.

A semiconductor head has not yet been practically applied due to the disadvantage of its slow speed from poor thermal response.

6.5.3 TEMPERATURE CONTROL

An important problem related to temperature control of the thermal head is heat hysteresis (Figure 6.6). For ideal temperature control, the thermal head temperature is controlled using a reference temperature when an input voltage signal is applied, and the thermal head temperature returns to its original value immediately when the input signal is set to zero. Due to heat hysteresis, this kind of control is not easily realized. However, several methods that make use of special arrangements have been developed.

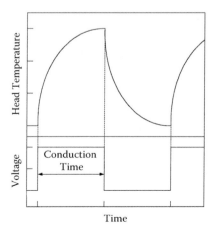

FIGURE 6.6 Transition of the surface temperature of a thermal head.

6.5.4 Concentrated Thermal-Transfer Head

Figure 6.7 shows a concentrated thermal head in which the pattern of the heating elements is used. The thermal head has many narrow regions, as shown in Figure 6.8. With a heating element of this type, high-temperature sections are created in the high-resistance section. Then at areas centered on these high-temperature sections, heat transfer starts. The transferred areas expand around these centering points as the applied energy is increased. The above is the major feature of this method. This method can also be used with melt ink sheets.

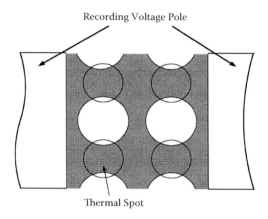

FIGURE 6.7 Concentrated thermal-transfer head.

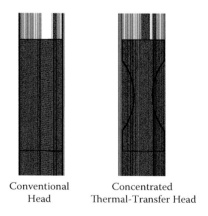

FIGURE 6.8 Heating elements of conventional and concentrated thermal heads.

6.6 VARIOUS IMPROVED PRINTER ENGINES

6.6.1 IMPROVEMENT IN PROCESSING SPEED

A one-pass full-color printer engine has been developed because thermal-transfer color printers are inferior to the electrophotographic printing system in relation to processing speed. Figure 6.9 shows the structure of the engine. The one-pass system has four independent recording sections: yellow (Y), magenta (M), cyan (C), and black (K). The recording sheet is passed to the four sections in order. The disadvantage of this system is that the sheet-forwarding section is complex and, therefore, color deviations tend to manifest easily. The speed of recording sheet varies due to the load and back tension during forwarding. This results in color deviations, even if the rotation speed of the drive roller is constant. By detecting the speed of the recording sheet directly using a detection roller and by controlling the drive roller to maintain a constant recording sheet speed, color deviations are prevented.

6.6.2 IMPROVEMENT OF DURABILITY OF THERMAL-TRANSFER PRINTER

In dye thermal-transfer, durability characteristics of the images are poor due to the inherent nature of the dyes themselves. For example, discoloration due to light exposure is a problem. To improve this problem, the following measures have been developed. In the recording layer, a compound is placed that reacts with a transferred dye and improves its stability. Also in the recording layer, an ultraviolet absorbing material or similar substance has been added. A transparent protective layer is sometimes applied on top for added stability.

6.7 OTHER PRINTERS BASED ON THERMAL TRANSFER

6.7.1 THERMAL RHEOGRAPHY

With this printer, a small hole is made at the center of the resistor in the thermal head and melted solid ink is transferred onto the recording sheets through the hole. This type of printer is based on the idea of inkjet recording.

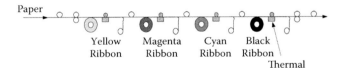

FIGURE 6.9 One-pass printer engine.

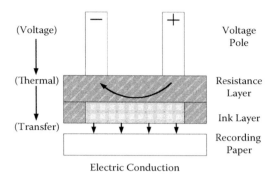

FIGURE 6.10 Structure of electrosensitive transfer.

6.7.2 ELECTROSENSITIVE TRANSFER PRINTER

Colored materials are transferred using joule heat from electric conduction through a coated layer of high electric resistance placed under the sublimable dye or thermal-transfer ink layer (Figure 6.10). This printer modulates the amount of ink transferred at each pixel in accordance with the duration of electric conduction. This printer is not popular because the production cost of recording materials for this system is high compared with that of other printer engines.

6.7.3 LIGHT-SENSITIVE MICROCAPSULE PRINTER

In place of an ink sheet, this system uses media with dispersed microcapsules. Capsule-containing sheets are exposed to ultraviolet light in an imagewise manner, creating a latent image. The ultraviolet exposure hardens capsules. The sheets are superimposed with recording sheets and put through a pressure roller. The pressure roller is able to crush only the unexposed capsules. The pigments in the crushed capsules are released and transferred onto the recording sheet.

The capsule wall is a polymer of urea and formaldehyde. Acryl monomers, polymerization initiator, spectral sensitizers, and leuco dyes are contained in the capsules. The recording sheet contains a developer. When released leuco dyes come into contact with the developer in the recording sheet, colors are produced.

6.7.4 LASER THERMAL-TRANSFER PRINTER

When higher resolution is desired, limitations in the thermal-head fabrication process become important because image resolution depends on the level of integration of heating elements in the thermal head. A new engine is being pursued to resolve this problem, i.e., a dye transfer printer using a laser as a heat source.

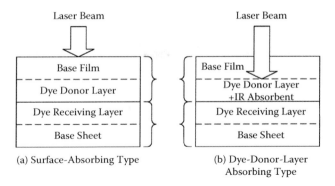

FIGURE 6.11 (a) Surface-absorbing printing media type and (b) dye-donor-layer-absorbing printing media type.

The dye sheets consist of a base film and a dye layer, and the recording sheets consist of a base sheet and a dye-receiving layer. This is basically the same as that used in conventional dye transfer printers. In addition, it is necessary to incorporate a layer that converts laser light to heat in the dye sheet. Figure 6.11 shows two types of printing media used for this printer. In contrast to the situation in Figure 6.11(a), in the construction of Figure 6.11(b), a layer that is infrared-ray absorbent and selectively absorbs semiconductor laser waves is included, and part of the laser beam is absorbed in the dye layer. By appropriately setting the parameters related to absorption, the amount of dye transferred can be increased.

A relatively high-power laser is required because the system is a heat mode. Previous research used a gas laser. In recent years, a high-power semiconductor laser has been developed. The semiconductor laser came from compact optical disc recording development.

In laser drawing, there are two types of image creation printer engines with respect to the recording surface on a drum. One is a printer engine in which the laser moves, and the other is a printer engine in which the laser from a fixed source is scanned using a mirror. For thermal-transfer recording with low recording sensitivity, the former systems are popular. Resolutions of 2540 dots per inch (dpi) or higher have been realized. As a countermeasure for the slow writing speed of the semiconductor laser system, multihead printer engines are being studied.

6.7.5 DIRECT THERMAL PRINTER

Coloring agents are applied in advance to the recording sheet. Color is formed when heat is applied locally onto the sheet using a thermal head. This system has advantages over others because of the possibility of high-speed printing.

6.8   CONTROLLER ASPECT

6.8.1   RESOLUTION

In regard to image resolution, 150 dpi (1980), 200 dpi (1987), 400 dpi (1996), and 600 dpi (1998) have been introduced. Greater than 1200 dpi has been available, but for limited purposes, such as color proofing.

6.8.2   TONE REPRODUCTION

The essence of sublimation thermal-transfer is that density modulation is possible for each dot. By an increase or decrease in joule heat, the density of each dot can be controlled. Heat levels are controlled by the pulse width of input signals. This control is an analog process.

The melt-type thermal-transfer printer produces solid and clear images. Thus, the printer is used in the preparation of bar codes. To achieve gradation, the dither method or another similar method is used, because the density characteristic of this recording method is binary. Thus, the resolution of this method tends to be low.

From the direct thermal printer, excellent coloring and image quality can be obtained, because it utilizes special sheets on which a coloring agent is applied.

As illustrated in Figure 6.12, the reproduction of color density by various printers can be classified into three methods. In the first method (bi-level pseudo density), the area of ink-covered dots changes; thus, when an image produced by this method is observed from a distance, a coherent density change can be observed. In the second method (multi-level pseudo density), an area is covered by a fixed number of dots whose color density is constant and whose dot size changes in relation to the density. The number of dots whose dot density and size are constant is changed in an area; this is called the bi-level area method. In the

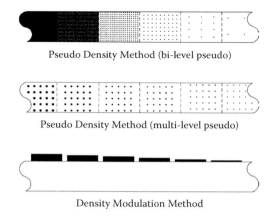

Pseudo Density Method (bi-level pseudo)

Pseudo Density Method (multi-level pseudo)

Density Modulation Method

FIGURE 6.12 Methods of the reproduction of color density.

third method, the density of each pixel changes. In this processing, each pixel can be given a continuous change of color density, and near-perfect gradation can be reproduced over the entire gradation range. In the first method, by incorporating a sharp heat-generating distribution in the narrow regions, area-based gradation, in which the transfer recording area within one pixel of the heating element is altered, has been realized. This combination method is called the dot-size variable multivalue area pseudo density reproduction method.

6.8.3 COLOR GAMUT

Sublimable dyes are suitable for use in full-color printer recording systems. The three primary colors, Y (yellow), M (magenta), C (cyan), of sublimable dyes have almost the same color gamut range as that realized by color offset printing.

6.9 USER ASPECTS

6.9.1 CHARACTERISTICS FOR APPLICATION

The dye sublimation printer is capable of producing color pictures whose image quality is comparable to that obtained by the photographic systems. Thus, when printing quality close to that of photographic images is required, such as in color proofing and production of pictorial color copies, this is suitable. The characteristics of the melt thermal-transfer printer offer sharp dot matrix images.

As for disadvantages, there are problems of poor durability, retransferring of transferred dyes, and early degradation of sections touched by fingers. Various methods to alleviate these problems have been developed for practical use. For example, sometimes the entire recording sheet is covered with a protective layer containing ultraviolet absorbents.

6.9.2 PRODUCT RANGE

There is a wide range of products using thermal-transfer technologies, from low-cost printers to high-end printers, from personal use to office use, from desktop publishing (DTP) use to color proofing use.

Small mobile printers for in-the-field use have been developed using thermal transfer technologies.

Instant photo printing at stores and home often depends on thermal-transfer technologies. The direct thermal printer is widely used in simple printers such as facsimiles and personal computers. In addition, this system is used in printers incorporated with measuring instruments, and high-end thermal-transfer printers are also used for color proofing in printing processes.

6.9.3 RUNNING COST

Although the running cost is high, many features make the printers cost effective for some applications. Because of the excellent color reproducibility, high density level (optical density value: 2.0), and high tone image (gradation: 64 or higher), they have been incorporated into products requiring high image quality, such as digital color proofing systems, video printers, DTP printers, and card-transferring machines.

For sublimation printers, multiple transferable ink sheets have been developed to reduce running cost. With these sheets, up to ten printing cycles with the same sheet are possible with no loss of density. In these ink sheets, large quantities of sublimable dyes are included in a thermoplastic resin layer.

6.10 STABILITY ISSUES

6.10.1 RECORDING DENSITY

Ambient temperature and heat accumulation in the thermal head can cause the thermal balance to be misdistributed, resulting in errors in recording density. These factors can cause problems of non-uniformity and problems in color reproducibility and resolution.

6.10.2 THERMAL ISSUES

There are two main thermal issues related to the thermal head. The first issue arises from the temperature increase of the thermal head. To maintain the operating temperature of the thermal head within limits that prevent destruction of either the head or the heat-resistant layer on the ink sheet, the head radiation fin is carefully designed. The second issue is related to the problem of constant changes in the thermal head temperature during printing. Measures counteracting transient temperature changes are required. These include optimum control by temperature detection using thermistors and temperature prediction using hysteresis data from gradation printing.

6.10.3 WEAR ISSUE

The thermal head wears due to friction because the thermal head maintains contact with the thermal-sensitive recording sheet as it moves. Head wear advances rapidly, in particular, if the thermal-sensitive recording sheet contains chemical compounds with high hardness. Chemical wear occurs in addition to this type of mechanical wear. It is corroded by alkali ions and other substances contained in the thermal sensitive recording sheets, because the surface of the thermal head is glassy.

6.11 CONCLUSIONS

There is a very wide range of products using thermal-transfer technologies, from low-cost printers to high-class printers, from personal use to office use, and from DTP use to color proofing use. The system has now been improved in terms of performance. The resolution of the thermal head has been improved from 600 to 1200 dpi. Further effort is being put forth toward high precision and low cost. The problem of image storage properties, which was the crucial issue for dye heat transfer, has been solved substantially by the stabilization of the dye through chemical reactions. Research continues in the search for better recording materials.

As described, there will be many improvements from the user's point of view and there are many expectations for dye transfer printers.

BIBLIOGRAPHY

T. Abe, Trends of thermography, ITE Technical Report, 11(26), 7–12, 1987.

S. Ando et al., A basic study of thermal transfer printing for improvement of print quality, *SID 85 Digest*, 160–163, 1985.

C.A. Bruce and J.T. Jacobs, Laser transfer of volatile dye, *J. App. Photo Eng.*, 3(1), 40–43, 1977.

H. Genno et al., Correction method of printed density with the sublimation dye transfer process, *SID Int. Symposium Digest*, 284–287, 1990.

T. Goto, Color reproduction of video printer, ITE Technical Report, 47(10), 1397–1400, 1993.

W. Grooks et al., Ribbon thermal printing, ribbon and head requirements, *IS&T, The 2nd International Congress on Advances in Non-Impact Printing Technologies*, p. 237, 1984.

K. Hanma et al., A color video printer with sublimation dye transfer method, *IEEE Trans.*, CE-31, 1985.

Y. Hori et al., Development of high definition video copy equipment, *IEEE Trans.*, CE-32, 1985.

M. Irie and T. Kitamura, High-definition thermal transfer printing using laser heating, *Journal of Imaging Science and Technology*, 37(3), 1993.

A. Iwamoto, Thermal printing technology, EID88–29, Japan, 1988.

T. Kanai, S. Hirahara, T. Ohno, K. Yamada, H. Nagato, and K. Higuchi, Digital–analog halftone rendition using ink-transfer thermal printer, 6th International Display Research Conference (Japan Display 86), 3, 1986.

S. Masuda et al., Color video printer, *IEEE, Trans. on Consumer Electronics*, CE-28 (3), 226–232, 1982.

N. Matsushiro, Dye transfer printing technology, *Encyclopedia of Imaging Science and Technology*, vol. 1, pp. 189–197, John Wiley, New York, 2002.

M. Mizutani and S. Ito, Thermal transfer printer, *Oki Technical Journal*, 51(2), 1984.

S. Nakaya, K. Murasugi, M. Kazama, and Y. Sekido, New thermal ink-transfer printing, *Proc. SID*, 23(1), 51–56, 1982.

I. Nose et al., A color thermal transfer printer with recoating mechanism, *SID 85 Digest*, 143–144, 1985.

H. Ohnishi et al., Thermal dissolution ink transfer for full-color printing, *IEEE Trans. on Electron Devices*, 40(1), 69–74, 1993.

K.S. Pennington and W. Crooks, Resistive ribbon thermal transfer printing, *SPSE Proc. of 2nd Int. Congress on Advances in NIP Technology*, 236, 1984.

O. Sahni et al., Thermal characterization of resistive ribbon printing, *SID 85 Digest*, 152, 1985.

M. Shiraish et al., Development of and A4-size color video printer, *SID 87, Digest*, 424–427, 1987.

N. Taguchi, H. Matsuda, T. Kawakami, and A. Imai, Dye transfer resistive sheet printing, *SPSE Proc. of the 4th Int. Congress on Advances in NIP Technology*, Thermograph Session, 532–543.

H. Tanaka, Multi contrast steps record of the thermal transfer printing method by wax ink, ITE Technical Report, 17(27), 19–24, 1993.

A. Tomotake et al., Structure/activity relationship of post-chelating azo dyes in thermal dye transfer printing system, *IS&T, The 14th International Congress on Advances in Non-Impact Printing Technologies*, 269–272, 1998.

K. Ttsuji, The trend of the research in thermal printing technology, EID89–38, Japan, 1989.

7 Film-Based Printers

*Tsutomu Kimura, Atsuhiro Doi, Toshiya Kojima,
Masahiro Kubo, and Akira Igarashi*

CONTENTS

7.1 HISTORY

Traditional photography is based on silver-halide chemistry. Historically, the ability to capture and keep photographic images became a reality in the 1830s when William Fox Talbot discovered a technique for preventing images from fading over short periods of time, and Louis Daguerre published his method for making daguerreotypes. Color film for consumers became available when Kodak's Kodachrome entered the market in 1935. As for instant photography, Polaroid Corporation put the first color film on the market in 1963, which was followed by the development of other types of instant photography. Throughout the years, photography has continuously improved. One important innovation was the processing method invented by 3M in 1964 that also led to the development of dry-silver color technology by 3M in 1986. Also in 1986, Fuji Photo Film announced the development of diffusion-transfer-type color thermal development silver-halide material technology, putting it on the market the next year.

Television technology made rapid progress following the beginning of practical broadcasting in 1935 (in Germany), when the technologies for handling an image as electronic signals and the displaying them on the cathode ray tube (CRT) advanced. In the late 1950s, during the space race, technologies for transmitting images, digital processing, and receiving their outputs developed. With such technological development, naturally, there was a growing need for printing images from electronic signals and digital data. Furthermore, when Sony introduced the Mavic digital camera in 1981, development of color printers for general use was accelerated remarkably.

In the course of this development, silver-halide photosensitive materials were naturally regarded as an advantageous means for making a hard copy of color electronic signals and digital images because of their high sensitivity and high image quality and also because of their diversified characteristics. There has been much technological research for the development of photographic materials designed for compatibility with a host of new recording systems. Various kinds of technological developments have been made for getting new materials for various recording systems matched with the exposure to light.

Exposing silver-halide photosensitive material with an image is known as recording the image. One common recording device is the CRT because of its relative compactness, and low cost CRTs are traditionally used to expose the entire image at once.

As a point light-source system, the drum-scanning system (halogen lamp, light emitting diode (LED), etc. as a light source) was developed early. In recent years a laser-scanning system has been developed, in which a laser beam is modulated to scan the exposure using a polygon mirror. As one-dimensional array exposure systems, there are LED array systems in which multiple LEDs are aligned, an electroluminescent display tube array system in which multiple electroluminescent tubes are aligned, and a light shutter array system in which PLZT (ceramic comprising Pb, La, Zr, and Ti) controls the polarized direction of penetrating light by applying a voltage. Another practical image recording system uses a two-dimensional array of digital micro-mirror devices (DMD) for controlling exposure (see Figure 7.1).

Corresponding silver-halide photosensitive materials are instant films, reversal films, color papers for direct recording, and positive prints after recording on negatives. It is possible to use non-specialized films that are used in standard photographic cameras, although in many cases their sensitivity to light is not appropriate for the exposure means of various systems. Technical developments have continued, including improvements in sensitivity, gradation, color-reproducibility range, and resolution of materials. Likewise, developments in exposure

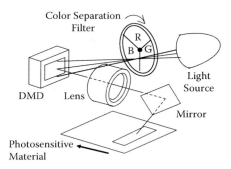

FIGURE 7.1 DMD exposure system.

means have contributed to changes in the light-emission spectrum, quantity of light, addressability, gradation control, and color-reproducibility.

In the following, we describe the exposure and printing technologies used widely by CRT and by fiber optical tube (FOT) as its modification. CRT exposure is a system in which a two-dimensional image displayed on the CRT screen is projected and exposed on a photosensitive material by optical means (Figure 7.2a). FOT exposure is a system in which the display area is a one-dimensional device of a flattened CRT and the image on the display and the photosensitive material are simultaneously moved for exposure while forming the image on the photosensitive material by optical means (Figure 7.2b and c).

When exposing a color image using a CRT, one usually exposes three times by applying three color filters to a black-and-white CRT successively (Figure 7.2a). Because no mechanical scanning system is necessary, the device is simple. Many video-printer products are on the market that are specially combined with instant photographic materials. Printers combined with color-paper systems were put on the market in the late 1980s. The FVP600 of Fuji Photo Film is an example (Figure 7.3).[1] The maximum recording width is 102 mm, and multiple exposures are executed using a high-brightness 7-in. CRT with 525 scanning lines, being able to produce an E-size print in an exposure time of approximately 3 seconds. The system used is shown in Figure 7.2, and the spectral sensitivities of the color-separation filter and color paper are shown in Figure 7.4. In the 1990s, large-size, high-performance devices were developed, and Agfa commercialized the AGFA DSP, which is a printer with a maximum width of 203 mm, utilizing a 9-in. flat-type CRT with 1024 scanning lines. It had a capacity of 1400 sheets per hour for L-size media and about 400 sheets per hour for a maximum size of 89 mm × 127 mm.

In color exposure systems using FOT, there are systems in which an image is created by applying three color filters to each of three sets of FOT and focusing them on a single point of the material using a mirror-lens system (Figure 7.2b),

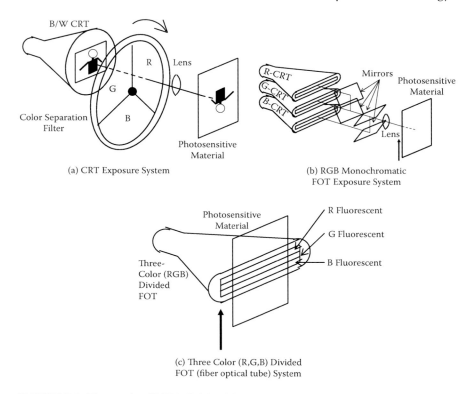

(a) CRT Exposure System

(b) RGB Monochromatic
FOT Exposure System

(c) Three Color (R,G,B) Divided
FOT (fiber optical tube) System

FIGURE 7.2 Three-color (RGB) divided FOT system.

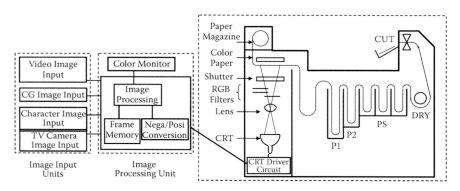

FIGURE 7.3 Diagram of an FVP600.

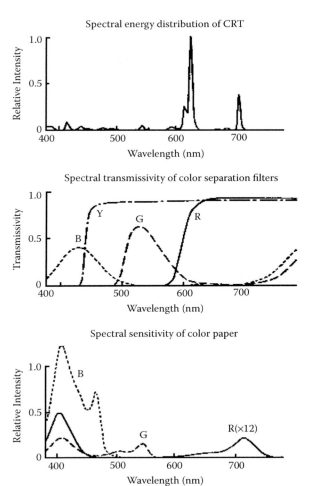

FIGURE 7.4 Spectral characteristics of FVP600 components.

or in which one FOT coated with three colors of electroluminescent materials (Figure 7.2c) is used. Both systems need an auxiliary scanning system to transport the material at a constant speed. The single FOT system is capable of exposing with the light source and the photosensitive material in contact so that the device can be downsized. The Konica VP100 was a commercialized example in which a single FOT system records on color paper. The first copy was obtained in 7 minutes and 30 seconds, and its capacity was 60 sheets per hour for the size of 130 × 180 mm.

7.2 PHOTO PRINTERS

Conventional mini-labs produce the image of a negative film on a silver-halide color paper using an illumination system, an optical system that focuses the image on color paper. After a certain exposure time, these are followed by development processing to get a print. At digital mini-labs, existing slides and prints may be copied. They are converted to digital data by scanning the originals using a CCD scanner. The digital data undergoes image processing, and the image is exposed on a silver-halide color paper, followed by development processing to get a print. Also, mini-labs record onto print paper digital data that come directly from digital still cameras (DSCs).

7.2.1 FILM SCANNING TECHNOLOGY

As for the CCD scanners used for reading images, there are line and area types, either of which may be selected on the basis of the required qualities for film scanning, mainly the number of pixels for reading and the reading speed.

7.2.1.1 Line CCD Image Reading Technology

Figure 7.5 shows the structure of the Fuji Film Frontier 350 film scanner using a line CCD. From the bottom, it consists of a light source for illumination, a film carrier, an image-forming optical system, and a CCD sensor for reading.

For illumination, a slit illumination optical system with a high light intensity has been developed to read a high-density negative film at a sufficient level of signal-to-noise ratio (S/N). As the light source, a halogen lamp with high stability was adopted, illuminating the slit illumination part in the film carrier through the cold filter, the light source iris mechanism, the negative and positive spectrum-compensating filters, and the diffusion box. Here, for illuminating a film with the high light intensity, a film-cooling mechanism has been adopted to prevent a rise in film temperature, making it a highly reliable and safe system.

The film carrier utilizes a line CCD to provide a mechanism for transporting film at a constant speed with high precision. A number of types are available for various film sizes.

Lenses for the image-forming optical system are sufficiently compensated for aberrations such as magnification chromatic errors, axial chromatic problems, and image distortion, and the optical magnification is adjustable in accordance with the necessary pixels for the output. Because of this, even for a large-size print over 10 in. × 12 in., it can secure sufficient reading resolution so that it may realize a sharpness equal to or better than that of analog printers and also provide optimal reading resolution for a wide range of negative sizes. Also, it can accurately read slide films that have a differential thickness or an uneven shape due to mount curling.

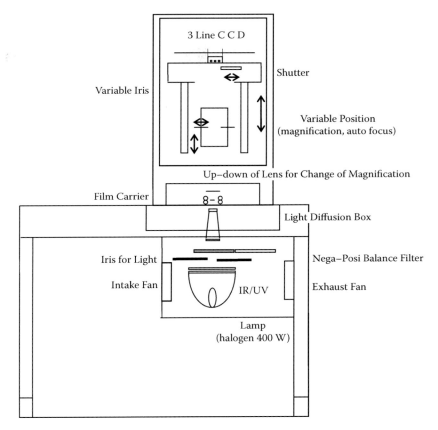

3 Line C C D

Shutter

Variable Iris

Variable Position
(magnification, auto focus)

Up–down of Lens for Change of Magnification

Film Carrier

8–8

Light Diffusion Box

Iris for Light

Nega–Posi Balance Filter

Intake Fan

IR/UV

Exhaust Fan

Lamp
(halogen 400 W)

FIGURE 7.5 Film scanner unit of a Frontier 350 (line CCD).

7.2.1.2 Area CCD Image Reading Technology

An area sensor has higher light utilization efficiency compared with a line sensor, and the film transport mechanism becomes simpler, which greatly contributes to miniaturization of the device. Figure 7.6 shows the structure of a film scanner for the Frontier 330, Fuji Photo Film, using the area CCD.[3]

The light source for the film scanner is LED (see Figure 7.7). LED is installed on a ceramic base with high thermal conductivity so that it has a high level of safety, an excellent characteristic of a film scanner. LED has excellent features such as lower power consumption than a halogen lamp, long life, its space-saving property, and negative film cooling is not needed. In particular, power consumption is a few watts, which is less than 1/100 of a halogen lamp, and, further, it lights up only at the time of scanning, contributing greatly to reduction in power consumption of the device. Problems with LED as the light source of a film scanner include securing a sufficiently high light intensity and controlling the

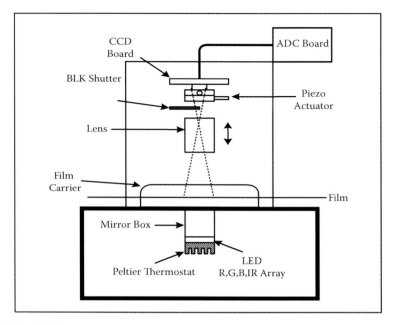

FIGURE 7.6 Film scanner unit of a Frontier 330 (area CCD).

FIGURE 7.7 (**See color insert following page 176.**) LED light source for a film scanner.

variations in light intensity and wavelength. The light intensity problem could be solved by using a honeycomb-type CCD, which will be described later. The variations in light intensity and wavelength could be suppressed by the use of a ceramic base and the temperature control by a Peltier element.

Figure 7.6 shows a diagram of a film scanner unit in which a Peltier element is used to stabilize the temperature of the light source. Figure 7.8 shows the light intensity of the light source before stabilization and after stabilization.

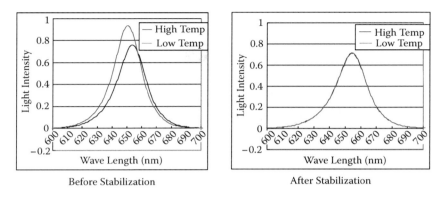

Before Stabilization   After Stabilization

FIGURE 7.8 Stability characteristics of light sources.

FIGURE 7.9 Honeycomb type of CCD.

Figure 7.9 shows a CCD developed exclusively for the Frontier 330. The effective pixel number is 3.2 million and it forms a honeycomb structure in which the pixels are arranged in a triangular arrangement. The opening shape is almost circular, and the area of the photoreceiver can be made large enough so that modulation transfer function (MTF) may have high isotropy and sensitivity to allow a wide dynamic range, an excellent characteristic for a film scanner. Also, to secure the necessary pixels for a large-size print, the Frontier 330 has two axes of a high-precision pixel-shifting mechanism using piezo. As shown in Figure 7.10, it can pick up an image, 1448 × 2172 pixels for each color with direct reading, 2048 × 3072 pixels with two readings having one shift between the readings, and 2896 × 4344 pixels for each color with four readings having three shifts between the readings. The maximum delivers more than four pages of about 12 million pixels. Further, infrared radiation (IR) is added to the illumination source of this device, allowing for a special channel that detects defects (Figure 7.11).

Before   After

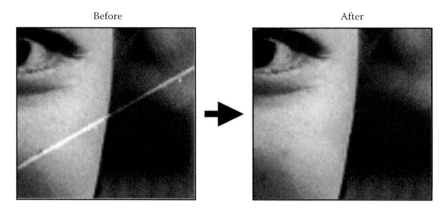

FIGURE 7.10 (See color insert following page 176.) Automatic scratch and dust restoration function.

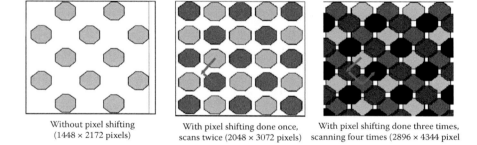

Without pixel shifting   With pixel shifting done once,   With pixel shifting done three times,
(1448 × 2172 pixels)   scans twice (2048 × 3072 pixels)   scanning four times (2896 × 4344 pixel

FIGURE 7.11 (See color insert following page 176.) Pixel-shifting method.

7.2.2 MARKING TECHNOLOGY

7.2.2.1 Exposure Technology

Among various exposure systems, several have been commercialized for digital mini-labs, and laser-scanning systems have become the mainstream because of their advantage in image quality. Those systems using a solid laser have the excellent characteristics of high image quality and the capability of miniaturizing the device. Figure 7.12 shows the printer structure of a Frontier 350 using a solid laser. Silver-halide paper is set in the form of a roll (two rolls in this device) and sheets cut to a specified length are transported upward, and after changing their courses to the horizontal direction, they are conveyed to the laser exposure part. The laser exposure part is transported at a high-precision speed.

The laser-scanning unit modulates each laser beam of RGB at a high speed with light-modulation elements, and the color paper being transported at high

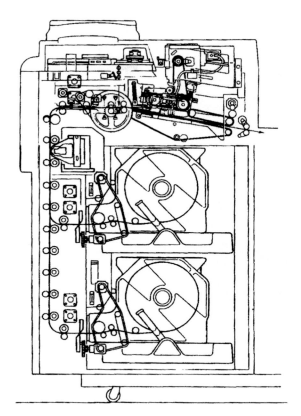

FIGURE 7.12 Laser exposure unit of Frontier 350.

speed is irradiated by these three laser beams for a constant-speed scanning exposure using a polygon mirror and fθ lens (Figure 7.13).

7.3   EMERGING TECHNOLOGY

7.3.1   PICTROGRAPHY

Photographic materials using silver-halides are superior to other color hard-copying materials in sensitivity and image quality. However, conventional silver-halide color paper has problems in the control of processing liquids used for development or processing speed. The diffusion-transfer-type color thermal development material technology (Pictrocolor System) is known by the name of Pictrography (PG).[4-5] It was introduced in 1987, and uses a small amount of water and heat, but no processing liquid.

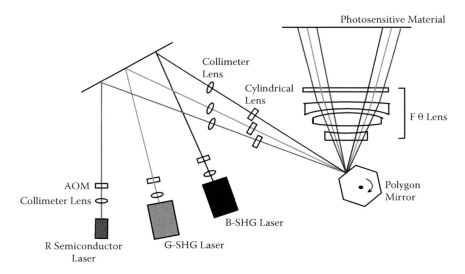

FIGURE 7.13 (**See color insert following page 176.**) Optical system for laser exposure.

7.3.1.1 Principle of Recording

A pictrography system is one kind of diffusion-transfer system, characterized by having all necessary development agents built in. Image formation in the Pictrocolor system proceeds as follows (see Figure 7.14):

1. Expose a donor film (thermal development photosensitive material)
2. Supply a small amount of water to the donor (approx. 0.7 cc/A4 or 10 cc/m^2)
3. Contact an image-receiving paper with the donor
4. Thermal development and dye transfer
5. Peel off the image-receiving paper from the donor

By exposing to light, a latent image (metallic silver nucleus comprising three to four silver atoms) is formed in the exposed silver-halide of the donor. Water supply, contacting, and heating generate an alkali, and development proceeds to release the water-soluble dyes. The released dyes diffuse and transfer to the image-receiving paper and are fixed by the mordant in the image-receiving paper. By peeling off the image-receiving paper from the donor, the residual water quickly evaporates due to the remaining heat, providing a print (Figure 7.15). The image-receiving paper thus obtained after printing contains only dyes and no unfixed silver-halide that would cause deterioration of the image quality.

7.3.1.2 Material Composition and Image Forming Process

The main composition of the donor is composed of hydrophilic binders such as gelatin, spectrally sensitized photosensitive silver-halides, dye-releasing redox

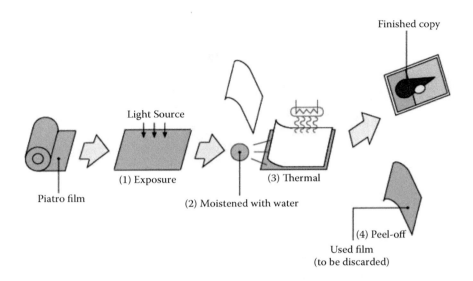

FIGURE 7.14 (See color insert following page 176.) Conceptual diagram of pictro-graphy.

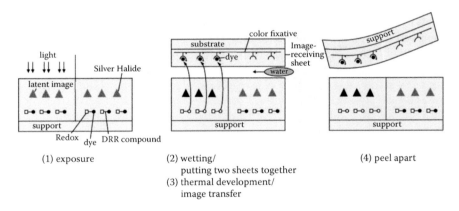

FIGURE 7.15 Process diagram of pictrography.

compounds (DRR compounds) known as dye material, and precursor basic metal compounds that generate alkali. The image-receiving paper consists of hydro-philic binders such as gelatin or various polymers, cationic mordant polymers for fixing dyes, and basic precursors that react with the basic metal compounds in the donor to form chelate compounds, generating alkali.

In the Pictrocolor system, the alkali necessary for advancing development is not generated until the donor and the image-receiving paper are brought into contact so as to better preserve the materials before use. For this reason, it is necessary for the donor and the image-receiving paper to be stable before contact. After contact, therefore, water-soluble agents contained in the image-receiving paper and the sparingly soluble basic metal compound react in a small amount of water to generate alkali in 2 to 3 seconds, even at room temperature. The alkali help to catalyze reactions in the DDR compound that releases diffusible dye (Figure 7.15).

7.3.1.3 PG Printer

Pictrography is the digital printer of the Pictrocolor system. The first-generation machine was developed in 1987, based on a drum exposure type with an LED light source. LED has excellent characteristics such as size, cost, dynamic range, light intensity, and the linearity between an electric current and a light intensity. However, the exposure system with LED has the following problems:

1. LED has a large luminous point so that an image-forming beam may not be narrowed down on a donor, which limits improvement in the recording density.
2. The drum exposure system has a limit in the revolution speed of the drum, hindering improvement in the recording speed.

Thus, it has been difficult to attain high resolution and high-speed recording.

With a change in the exposure system to LD in 1993, the donor was also improved to get much better image quality. As compared with LED, LD is easy for narrowing the beam diameter, being suitable for high-density recording, and replacement of the exposure system from LED to LD has improved resolution from 284 to 400 dpi. Also, the convergence efficiency of the collimeter lens is so high that exposure light intensity may be enough for high-speed recording. Furthermore, by combining LD with a polarizer consisting of a polygon and an fθ lens, photosensitive material may be exposed in the course of transport in a plane, which makes it possible to miniaturize the device (Figure 7.16).

On the other hand, the LD exposure system has the following defects:

1. The practical range of LD wavelength is limited.
2. The electric current and optical output characteristics of LD vary significantly with temperature change.
3. When LD light beams of three wavelengths form images with the same lens, the wavelength dependence of image-forming characteristics such as color aberration appears as a color shift, variations in the beam diameter, etc.
4. When the optical parts have transmittance and reflectance distributions in the main scanning direction, it causes shading. Particularly when the shading characteristics are different among colors, differences become discernible as color shading.

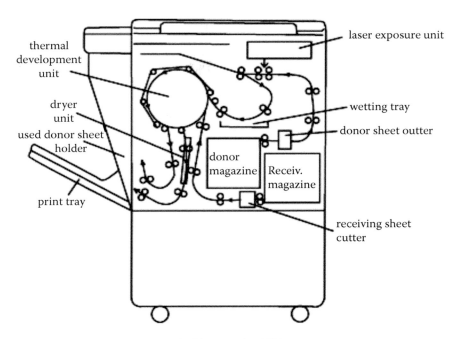

FIGURE 7.16 Internal diagram of a Pictrography 3000.

7.3.1.4 High-Quality Imaging Technology

LD wavelengths used in pictrography are 680, 759, 810 nm. To prevent unintentional color mixing even in this narrow wavelength range, it is necessary to design the donor spectral sensitivity distribution so that there would be sharp, non-overlapping peaks in this wavelength range. On the material side, it has become practical for the first time because of J-band sensitization technology that regularly arranges molecules of sensitizing dyes on a surface of the silver-halide crystals for a photosensitive layer to have a peak at 750 nm. For the photosensitive layer having a peak at 810 nm, multi-functional ultra-red dye was introduced with a sharp absorption at 750 nm. With the introduction of these technologies, spectral sensitivities with excellent color separation were realized (Figure 7.17).

LD exposure is a pulse-width modulation system with a fixed light intensity. The modulation is controlled by 12 bits having a sufficient density resolution. For variations due to differences in LD exposure parts and sensitivity variations of photosensitive materials, the printer is provided with a calibration system in which the built-in densitometer measures the relation between the exposure control signal and the print density to allocate the optimal 12-bit signal to an 8-bit image signal.

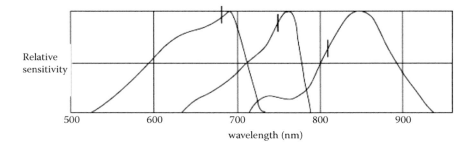

FIGURE 7.17 Spectral sensitivity of pictrography media for LD exposure.

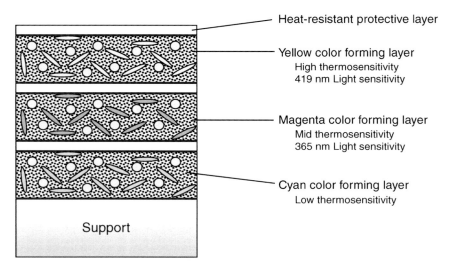

FIGURE 7.18 (**See color insert following page 176.**) Basic structure of TA paper.

7.3.2 THERMO-AUTOCHROME METHOD

7.3.2.1 Recording Principle

Until the 1980s, it was believed that full-color printing by a direct thermal print system was possible in principle but not in practice. In 1994, however, Fuji Photo Film commercialized a direct thermal printer based on a thermo autochrome (TA) system.[6-8] Thermo autochrome is a coined term meaning a system in which all the necessary mechanisms for color printing are incorporated into recording paper, and by repeat heating and light exposure (automatically), one may get a color print.

Figure 7.18 shows the cross-sectional structure of TA paper. On a substrate, each thermo-sensitive layer that renders cyan, magenta, or yellow color is successively coated, and a heat-resistant protective layer is provided as the top

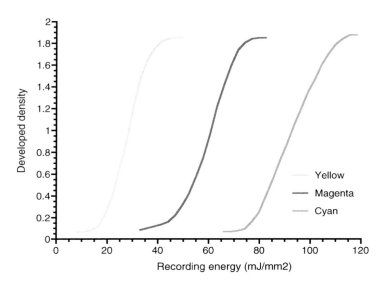

FIGURE 7.19 Thermal recording characteristics of TA paper. (The density is measured with an X-rite reflective densitometer. The recording energy is calculated from the electric energy supplied to the thermal head.)

layer. Each layer reacts with a different thermal energy to develop the color. The uppermost yellow layer reacts with the lowest thermal energy, and the lowermost cyan layer reacts with the highest. Figure 7.19 shows the thermo-color development characteristics. The two upper layers of magenta and yellow have thermo-sensitivities and photosensitivities as well. In the magenta-developing layer, the incorporated color-developing component is decomposed by 365-nm ultraviolet rays and loses color-developing capacity by heating. Accordingly, after forming an image by heating, the image may be fixed by exposing the total paper to 365-nm ultraviolet rays. Similarly, the yellow-color developing layer loses color-developing capacity by 419-nm ultraviolet rays.

By building such a mechanism into the recording paper, full-color printing becomes possible by the following simple procedures:

1. Apply low thermal energy corresponding to yellow image information to a TA paper to record a yellow image
2. Expose it to 419-nm ultraviolet rays to fix the yellow image
3. Apply an intermediate thermal energy corresponding to magenta image information to record a magenta image
4. Expose it to 365-nm ultraviolet rays to fix the magenta image
5. Apply high thermal energy corresponding to cyan image information to record the cyan image

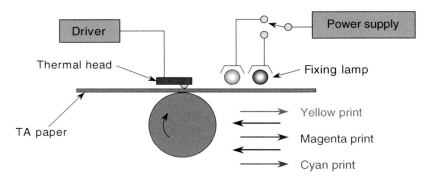

FIGURE 7.20 Basic configuration of a TA printer.

Figure 7.20 shows the basic configuration of a TA printer. The printer is composed of a thermal head for thermal recording, one ultraviolet-ray fluorescent lamps of 419 nm and one of 365 nm, and platen rolls for driving. The TA paper is fed from the left side of the thermal head. In the first process, the yellow image is simultaneously recorded and fixed. That is, the yellow image is recorded by the thermal head and, at the same time, the image is fixed by the 419-nm ultraviolet-ray lamp placed downstream. Next, the paper is returned to its original position to record and fix the magenta image. Last, the cyan image is recorded. Thus, by having the paper move back and forth two and a half times, one can complete full-color printing.

The merit of the TA system is that TA paper is the only material necessary for printing, and it does not require any consumables such as ink or toner at all; accordingly, there is no waste disposal associated with printing. Further, the drive is only for transporting the paper, so one can build a simple and highly reliable system.

7.3.2.2 TA Paper

Because a yellow image layer recorded by low thermal energy may be developed by an intermediate energy for magenta image recording if left as it is, it is necessary to fix the yellow image (lose its thermal sensitivity) by some means. In TA paper, this problem has been solved by introducing a diazonium-salt compound as a color-developing material.

The diazonium salt compound reacts with a coupler to form a dye. It is decomposed by the light corresponding to its absorption wavelength, losing its reactivity with the coupler. In the recording layer, a diazonium salt compound and the coupler are dispersed. They do not interact until they are heated. When the layer is exposed to light, the thermally recorded image is fixed. Figure 7.21 shows the absorption spectrum of the diazonium salt compounds used for the magenta color-developing layer and the yellow color-development layer. By making them have different spectral absorption characteristics, only the yellow image may be fixed selectively when one exposes it with 419-nm ultraviolet rays in

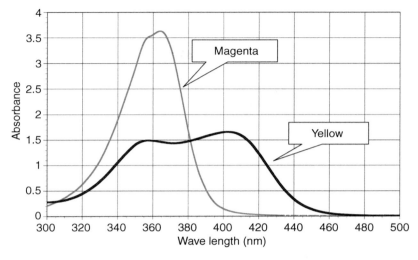

FIGURE 7.21 Spectroscopic photosensitivity of diazonium salt compounds. (Measurements were made on methanol solutions in concentrations of 25 mg/L.)

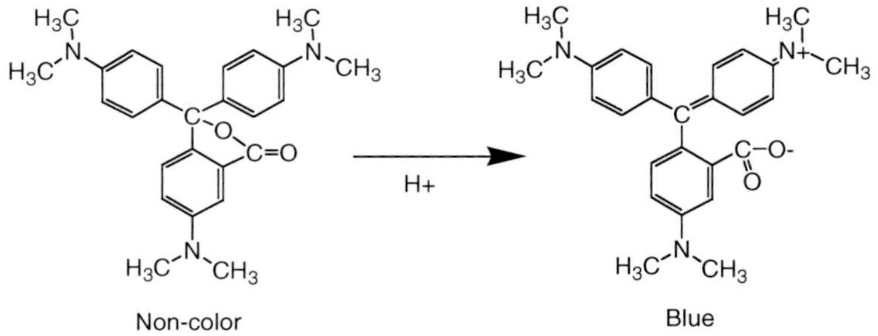

FIGURE 7.22 Color formation reaction of a typical leuco dye.

advance. The magenta image is fixed by 365-nm ultraviolet rays because the yellow image has already been fixed.

Because the cyan color-developing layer does not require fixation, a leuco dye and an organic acid used for conventional heat-sensitive recording paper are used as color-developing materials. The leuco dye is a kind of pH-indicating agent that is colorless in the basic and neutral ranges but forms a dye under acidic conditions (Figure 7.22). This reaction is reversible.

7.3.2.3 Heat-Responsive Microcapsule

In the color-developing layers of TA paper, two colorless compounds that form a dye through a reaction are included separately in the recording layer, and they react by heating to form a dye image. Thus, it is important to have a system in which the two compounds are stably separated at room temperature and both react rapidly by heating.

Heat-responsive microcapsules have been introduced into TA paper. Microcapsules are minute vessels of core–shell structure. The core components are protected from the surroundings by the shell, and with a heat-responsive microcapsule, permeability of substances through a polymer membrane constituting the shell varies significantly with the surrounding temperature. More concretely, the shell comprises a poly(urea-urethane) membrane. Below glass transition temperature (T_g), it has a very low permeability of substances, and above T_g, the permeability increases by severalfold. This change is probably due to a drastic change in the force of hydrogen bonding within molecules or among molecules of the membrane-forming polymer around T_g.

In the magenta and yellow color-developing layers of TA paper, microcapsules consist of a core where a hydrophobic diazonium salt compound is dissolved in a hydrophobic and high-boiling-point solvent and a shell of poly(urea-urethane) membrane coexists with a coupler and an organic base. At normal temperature (below the T_g of the shell membrane), the diazonium salt compound is insulated from outside under hydrophobic conditions so that it is quite stable despite its high activity. However, when heated, the three components are mixed to generate a dye-forming reaction.

With the cyan color-developing layer, Leuco dye is similarly included in the heat-responsive microcapsule.

7.3.2.4 TA Printer

As shown in Figure 7.20, key parts of a TA printer consist of a thermal head, a set of ultraviolet fluorescent lamps, and mechanical parts for paper transport.

The thermal head consists of minute heating elements arranged on a ceramic base, which can thermally record 300 to 600 dpi. Thermal energy applied to TA paper may be controlled by the magnitude of the electrical energy to the heating elements; however, the heating temperature varies, depending on the surrounding temperature or the base temperature of the thermal head, even if the same electrical energy is applied. Thus various controls are employed so that stable recording density may be reproduced, irrespective of the printing environment. Further, as is apparent from the thermal color-development characteristics in Figure 7.19, the magenta color development starts in the saturated density region of yellow, and the cyan color development starts in the saturated density region of magenta. Thus, a high-density yellow or magenta is likely to cause mixed colors. To prevent this, a three-dimensional look-up table (LUT) is installed in the printer so optimal heating conditions of yellow, magenta, and cyan (YMC) may be chosen for RGB image information. More concretely, when saturation is demanded, the density to prevent mixed colors is suppressed, and when more density than saturation is required, the color is heated sufficiently to get a high density.

Figure 7.23 shows the emission spectra of ultraviolet fluorescent lamps for fixation. The ultraviolet fluorescent lamps are inexpensive and highly efficient;

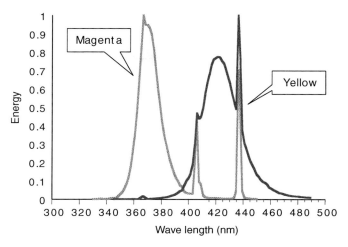

FIGURE 7.23 Emission spectra of the fixing lamps.

FIGURE 7.24 Digital color home printer using a TA system (Fuji CX-400).

however, they have a defect: illumination energy changes easily with the surrounding temperature or deterioration. A TA printer is designed so that printing may be carried out under optimal conditions by always monitoring the light intensities. Figure 7.24 shows a digital home printer for directly obtaining full-color prints from a digital camera output. At a speed of 20 sheets per hour, printing of L-size (89 × 127 mm) without edge margins is possible, with the equivalent image quality of a conventional silver-halide photo. Figure 7.25 shows a TA printer in which thermal heads for YMC recording are separated individually.

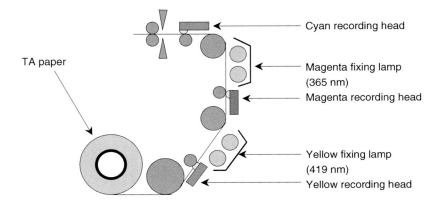

FIGURE 7.25 (**See color insert following page 176.**) Configuration of high-speed three-head tandem digital color printer.

7.3.2.5 Summary

TA is a simple and highly reliable full-color printing system. It provides prints equivalent to silver-halide photos in a completely dry system, without generating any waste materials. For the present, it is used for printing from digital cameras that require instant output, but in the future its use may be expanded to include applications in the medical field and various other markets.

Technically, because all the necessary components for color development are included in the paper, it is necessary to improve the whiteness at the image base; however, the storage level of whiteness equivalent to that of silver-halide photos has been attained. As for the durability of printed images, it is still inferior to that of silver-halide photos; however, it has reached the level of several tens of years in dark storage and is being improved year by year.[9]

As for the printer, performance improvements in ultraviolet-ray LEDs will allow its use as an illumination light source for fixation in the near future. The illumination energy of LEDs is far more stable than that of fluorescent lamps, and it has a longer life and compactness. It will enable a system design in which ease, simplicity, and high reliability — merits of TA — will become more conspicuous.

REFERENCES

1. K. Shiota, Optical printers, *Journal of Institute of Image Information and Television Engineers*, 43, 1223–1229, 1989.
2. Y. Ozawa et al., Development of Digital Minilab System Frontier 350/370, Fuji Film Research & Development, 45, 45–41, 2000.
3. Y. Nakamura et al., Development of Digital Minilab System Frontier 330, Fuji Film Research & Develoment, 48, 15–21, 2003.

4. M. Kubo, A. Ueshima, and M. Okino, Full-Color Laser Printer, Pictrography 3000(1), Hardware, Technical Research Report, Institute of Electronic Information Communication, Vol. 93, No. 274, Oct. 1993.

5. Kamosaki, Yokokawa, and Inagaki, Full-Color Laser Printer, Pictrography 3000(2), Photosensitive Materials, Technical Research Report, Institute of Electronic Information Communication, Vol. 93, No. 274, Oct. 1993.

6. A. Igarashi, T. Usami, and S. Ishige, IS&T's 10th International Congress on Advances in NIP Tech., pp. 323–326, 1994.

7. M. Sato, M. Takayama, and M. Tsugita, IS&T's 10th International Congress on Advances in NIP Tech., pp. 326–329, 1994.

8. A. Igarashi and T. Usami, *J. Inf. Recording*, 22, 347–357, 1996.

9. K. Minami, S. Sano, A. Igarashi, *Japan Hard Copy*, 98, 91–94, 1998.

Part III

The Management of Color

8 Color Management

Mitchell R. Rosen

CONTENTS

8.1 INTRODUCTION

Color reproduction workflow involves a pair of color devices: a source device for capturing or creating a complex original stimulus and a destination device for re-creating the original's appearance. Image processing steps enhance, maintain, or degrade the appearance of the image or prepare it for rendering on the destination device. In traditional analog workflows such as film photography, the processing steps are usually chemically based. Modern workflows such as digital photography often have processing steps implemented in software or firmware algorithms.

Color management is designed for digital workflows to provide specialized color processing between the capture and rendering stages with the goal of maintaining color appearance. The original can be found in the real world or it can derive from a color device. The original sometimes is of unclear pedigree. Table 8.1 describes all the possible source/destination color reproduction combinations.

Table 8.1 can be deceptive, as it makes the problem of color reproduction seem easy. The source column of the table lists only cameras, scanners, displays, printers, and files. The destination column mentions only displays and printers. If the problem truly boiled down to only five possible inputs and two possible outputs, as implied by the table, then color management should be uncomplicated.

It turns out that within each category of device, there are many different technologies that could potentially be involved. Each technology introduces different color reproduction characteristics. Also, within a family of similar technologies, there are many design choices that go into differentiating the various devices. These, too, have a large impact on how color devices respond. Color management, when practiced well, makes it possible for a single reproduction system to allow any number of different device types to participate.

TABLE 8.1
Color Reproduction Input/Output Device Combinations

Where Original Is Found	How Original Is Brought into the Color Reproduction Workflow (source)	Where Reproduction Is Rendered (destination)
In the real world	Camera	Display
		Printer
	Scanner	Display
		Printer
On a color device	Display	Display
		Printer
	Printer	Display
		Printer
Ill-defined	File	Display
		Printer

For color management to work, the stimulus and response character of a digital device must be characterized. Digital cameras are stimulated by color and respond with digital values. Scanners are similar. Displays and printers are the opposite: they are stimulated by digital values and respond with colors. Table 8.2 summarizes this.

There are two basic approaches to color management. Each requires knowledge of the stimulus/response character of color devices. The two approaches differ in how they take advantage of device characterizations. One approach to color management calibrates every device to act like a particular, standard pseudo-device. One such popular pseudo-device is described by the sRGB standard (see Figure 8.1).

TABLE 8.2
Stimulus/Response Pairings for Digital Devices

Digital Device	Stimulus	Response
Camera	Color	Digital Values
Scanner	Color	Digital Values
Display	Digital Values	Color
Printer	Digital Values	Color

FIGURE 8.1 sRGB calibration-based approach to color management. Source devices are calibrated to deliver standard sRGB digital values as their "response" and destination devices are calibrated to accept sRGB digital values as their "stimulus."

The second color management approach is profile based. Here device characterizations are packaged into standard data structures known as profiles. Two profiles will be used at the time of color reproduction. The first profile describes the source device and the second profile describes the destination device. Color processing will use these profiles to attempt to maintain the color appearance of the original.

Each device has its own profile. The profile teaches the color management system how to relate the device's digital values to colorimetry. Instead of emulating a standard pseudo-device, as in the sRGB case, here each device is free to act in its native way. Figure 8.2 illustrates this approach. The International Color Consoritum (ICC) defines the industry standard for profile-based color management.[2]

This chapter concentrates on the profile-based approach to color management as defined by the ICC. Because this approach allows devices to retain their intrinsic stimulus and response character, it has the potential of maximizing quality.

The ICC profile specification determines the format of an ICC profile. Basically, the profile file is a tagged structure much as the TIFF image file format is a tagged structure. Such a format includes a standard header, a table of contents,

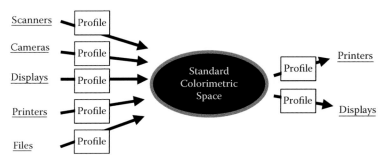

FIGURE 8.2 ICC profile-based approach to color management. Every device has its own profile that describes that device's relationships among digital values and color.

and a series of data structures called tags that are enumerated in the table of contents.

An ICC source device profile contains tags that describe the relationship between input digits and colorimetry. Tags in an ICC destination profile describe the relationship between colorimetry and output digital values. The most popular operating systems have implemented support for ICC profiles. These technologies have been implemented in Windows as image color management (ICM) and in Macintosh as ColorSync.

8.2 ICC COLOR MANAGEMENT GENERAL APPROACH

Figure 8.3 describes a standard profile-based color management approach. An image is captured by or created on a source device. The profile describing the source device and the profile describing the destination device are used to guide image processing. Afterward, the processed image is rendered on the destination device.

Equation 8.1 and Equation 8.2 illustrate one of the typical transformations from device digit to device independent color that can be encoded within an source profile. Equation 8.1 uses one-dimensional look-up tables (1D LUTs) that linearize an input RGB signal. Equation 8.2 follows with a 3×3 matrix that transforms the linear values to XYZ.

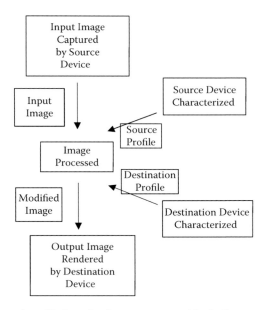

FIGURE 8.3 General profile-based color management block diagram.

$$[R'] = LUT_{R \to R'}[R]$$

$$[G'] = LUT_{G \to G'}[G] \tag{8.1}$$

$$[B'] = LUT_{B \to B'}[B]$$

$$\begin{bmatrix} X \\ Y \\ Z \end{bmatrix} = \begin{bmatrix} a_{11} & a_{12} & a_{13} \\ a_{21} & a_{22} & a_{23} \\ a_{31} & a_{23} & a_{33} \end{bmatrix} \begin{bmatrix} R' \\ G' \\ B' \end{bmatrix} \tag{8.2}$$

where RGB are the digital values of the source device, a_{ji} are the matrix coefficients to estimate colorimetry, and XYZ are the tristimulus values for an object under a given illuminant.

For a common transformation chain often parameterized by a destination profile, Equation 8.3 through Equation 8.5 are presented below. In Equation 8.3, L*a*b* values are modified through individual 1D LUTs. A multidimensional LUT in Equation 8.4 transforms the modified L*a*b* values to linearized cyan, magenta, yellow, black (CMYK). A final set of 1D LUTS in Equation 8.5 is applied to appropriately shape the CMYK signals for the destination printer.

$$\left[L^{*'} \right] = LUT_{L \to L'} \left[L^{*} \right]$$

$$\left[a^{*'} \right] = LUT_{a \to a'} \left[a^{*} \right] \tag{8.3}$$

$$\left[b^{*'} \right] = LUT_{b \to b'} \left[b^{*} \right]$$

$$\begin{bmatrix} C' \\ M' \\ Y' \\ K' \end{bmatrix} = LUT_{Lab' \to CMYK'} \left[L^{*'}, a^{*'}, b^{*'} \right] \tag{8.4}$$

$$[C] = LUT_{C' \to C}[C']$$

$$[M] = LUT_{M' \to M}[M']$$

$$[Y] = LUT_{Y' \to Y}[Y'] \tag{8.5}$$

$$[K] = LUT_{K' \to K}[K']$$

To use profiles, the image-processing stage of the color reproduction block diagram must be able to make appropriate use of the transformation coefficients found in the tags of the input and output profiles. The actual transformation chains implied by the profiles may be directly applied to images or they may be concatenated to form more efficient processes. In addition to being able to perform the color transformations described by the profiles, an image-processing engine must also be able to convert among ICC's supported device-independent color-spaces known as profile connection spaces (PCS). Currently supported PCSs are variants of XYZ and L*a*b*.

If a color reproduction system were being used to convert an image from an input device with a profile utilizing Equations 8.1 and 8.2 to an output device with a profile utilizing Equation 8.3 through Equation 8.5, then the image-processing stage would need to provide an additional transform from XYZ to L*a*b* between Equation 8.2 and Equation 8.3.

8.3   ICC COLOR MANAGEMENT

The ICC requires that tags describe device characteristics in terms of D50 colorimetry. Chromatic adaptation must be applied for measurements taken under different conditions. Also, a special normalization of the colorimetric values is specified. This normalization is convenient for processing images to be rendered as reflection prints because it guarantees that very bright whites on input will be mapped to paper white on output. In ICC terminology, this manipulated colorimetry is called profile connection space (PCS).

For most color capture devices, the mapping from its three channels and to three dimensions of colorimetry is overall one-to-one. Where the actual mapping is many-to-one or one-to-many, this is due to differences between the instrumental metamerism of the input device and the metameric characteristics of the standard observer. For most printing devices, there is much redundancy between colorimetry and output digits due to the typical presence of a fourth colorant. Makers of ICC profiles must deal with these conflicts and provide color transformations with single mappings in each instance.

Gamut mapping is an important concept in color reproduction. The goal of color reproduction is to maintain the color appearance of an original. Sometimes this means attempting to match exactly the colorimetry of the original. Other times, color appearance actually is improved when moving between two different media by using different colorimetry on the reproduction. Either way, an algorithm determines what color should be printed or displayed. If that color is not achievable on the output device, gamut mapping is necessary.

Figure 8.4 is a gamut-mapping cartoon. The contour indicates the gamut limit of a device and the star is considered an out-of-gamut color that needs to be reproduced. There are many choices available to the profile-builder as to how to map colors into the gamut. One popular method is to choose a particular point on the L* axis (a centroid) and map all colors toward it. Figure 8.4b illustrates such an approach.

Regardless of how the out-of-gamut color is brought into the device gamut, a second question is whether the point should be mapped to only the gamut surface or brought further into the gamut. One reason to bring colors further

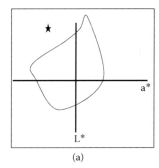
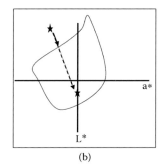

(a)   (b)

FIGURE 8.4 Gamut-mapping cartoon. Contour indicates gamut limit of a device; a star indicates out-of-gamut value to be mapped. For this particular gamut-mapping scheme, a second star on the L* axis indicates a centroid value toward which out-of-gamut colors are mapped.

in-gamut is to allow for differentiation among out-of-gamut colors. Bringing some out-of-gamut colors into the bulk of the gamut is known as creating a soft-tuck vs. the idea of a hard-tuck to the gamut surface.

Creating a soft-tuck becomes a very complicated process because in-gamut colors must also be compressed toward the center of the gamut. Figure 8.5 illustrates this point. Figure 8.5a shows the compression of the in-gamut colors to make room for some out-of-gamut colors.

Figure 8.5b shows four gamut-related regions that are sometimes defined. The central core near the neutral axis is sometimes left alone without any compression. The soft region between the central core and the gamut surface is often compressed, moving in-gamut colors and making room for some out-of-gamut colors. Then there is an out-of-gamut region that defines which colors are brought into the gamut bulk. Finally, there is the outermost area from which all colors are mapped directly onto the gamut surface.

An entity that applies ICC-compliant image processing is called a color management module (CMM). The CMM processing typically involves a pair of profiles, one referring to the image-capture device and the other referring to the image-rendering device, as shown in Figure 8.6a. Common image transformation chains that may be encoded in source and destination profiles were discussed in Section 8.2. Equation 8.1 and Equation 8.2 describe an image transformation chain typically encoded in the tags of a source profile and Equation 8.3 through Equation 8.5 are typically parameterized within the tags of a destination profile.

Before processing images, the CMM will often apply data processing to the profile data for the sake of efficiency. The fact that each ICC profile is tied to the common PCS makes the concatenation of a series of transforms fairly straight-forward and low-cost computationally. Equation 8.6 through Equation 8.8 show one possible transformation chain that could result from collapsing portions of the full set of Equation 8.1 through Equation 8.5. These equations demonstrate a popular image-processing chain that starts with one-dimensional LUTs applied

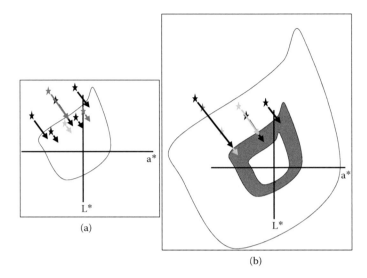

FIGURE 8.5 Soft-tuck cartoon. Out-of-gamut colors are brought into the bulk of the gamut. Some in-gamut colors need to be compressed to make room.

to three input channels followed by a three-dimensional look-up and possibly a final one-dimensional look-up applied to each output channel (see Figure 8.6b). Computationally and with respect to memory requirements, the image-processing chain described here is very feasible.

Equation 8.6 is the same as Equation 8.1. Equation 8.2 through Equation 8.4 are concatenated into the multi-dimensional LUT of Equation 8.7. Equation 8.8 is no different from Equation 8.5. Note that in building this more efficient color-processing chain, it is still necessary for the concatenation engine to plug in an extra transformation that resolves the XYZ and L*a*b* discontinuity found between Equation 8.2 and Equation 8.3. That additional step is also captured in the LUT of Equation 8.7.

$$[R'] = LUT_{R \to R'}[R]$$

$$[G'] = LUT_{G \to G'}[G] \tag{8.6}$$

$$[B'] = LUT_{B \to B'}[B]$$

$$\begin{bmatrix} C' \\ M' \\ Y' \\ K' \end{bmatrix} = LUT_{RGB' \to CMYK'}[R', G', B'] \tag{8.7}$$

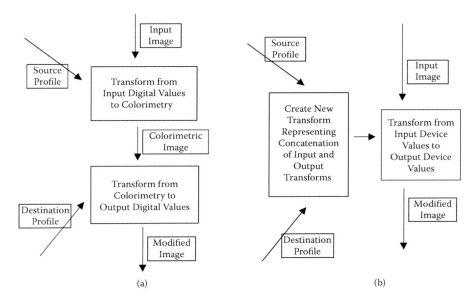

FIGURE 8.6 (a) Logical ICC image-processing block diagram. (b) Typical CMM image processing with efficiencies. Transform box may include 1-D and 3-D look-ups, matrices, and other functional forms.

$$[C] = LUT_{C' \to C} [C']$$

$$[M] = LUT_{M' \to M} [M']$$

$$[Y] = LUT_{Y' \to Y} [Y']$$

$$[K] = LUT_{K' \to K} [K']$$

(8.8)

8.4 HISTORY OF ICC

For a number of years prior to the first ICC specification in 1994,[3] a few companies tried to promote and sell their own color management solutions including Kodak, Tektronix, and EFI. These tended to involve special software that a user would manually invoke to process images. The packages were commercial flops.

One company, Adobe, did have a successful color management launch, which was implemented in every rasterizing image processor (RIP) that supported the PostScript Level 2 Language.[4] Today's ICC approach is remarkably similar to what was brought to the masses by Adobe in 1990. However, Adobe's color management was limited to only Postscript printers and, thus, did not offer a general solution.

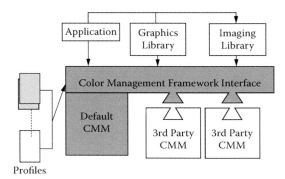

FIGURE 8.7 An operating system's view of color management showing how color-naive applications and libraries can use color management, and how those with added-color-value can plug-in. (Adapted from ICC specification.[2])

At least one ad hoc industry group, the Association of Color Developers (ACD), was formed during those pre-ICC days to come up with an open standard to bootstrap the acceptance of universal color management. The effort went nowhere. The ACD did create one lasting legacy, however. It put together a schematic that looks much like the one in Figure 8.7. The group realized that selling color management was going to succeed only when the operating system vendors understood how they could facilitate better color reproduction and when applications vendors without color expertise were supplied with an approach that would not burden them with the development costs of creating color smarts.

Illustrations like the one in Figure 8.7 were designed to speak to the operating system and application provider community. To a significant extent, this sketch was effective in selling the idea. What was needed at that point was a leader to take the plunge and implement the infrastructure. That leader came in the form of Apple Computer. Apple made the investment and in January 1993 introduced an extension to its operating system toolbox: ColorSync.

The original ColorSync printer model had limited value. It was based on a transform that was only accurate for perfect dot-on-dot printers.[5] Apple realized early on that ColorSync was going to need significant industry help and buy-in to become a success. Gerry Murch, who led Tektronix's early color management efforts, was then at Apple and headed the drive to create industry consensus. He brought together a group that, at first, had the name ColorSync 2 Consortium. At that time it was heavily centered on helping Apple define the next level of ColorSync. Kodak, Adobe, and others already had fully functioning color management systems that were incompatible with the original ColorSync. Those companies had motivation to see a more mature ColorSync that supported their approaches. Device manufacturers and graphics arts applications providers were also motivated to see Apple succeed so that they did not have to implement or buy their own color management technology. Hewlett Packard, with extreme dominance in the desktop- and office-printer market at the time, did not join the consortium until close to its roll-out and could easily have derailed it if the company had been motivated.

As the movement proceeded, the name and mandate of the consortium changed in order to encourage companies with no reason to help Apple to consider joining. From ColorSync 2 Consortium, the next name was InterColor and eventually the International Color Consortium (ICC). As evidence of the successful detachment of the consortium from Apple, even Microsoft eventually joined the consortium.

In the early formation of ColorSync 2, Adobe was very successful in helping steer the consortium in several important ways. As already mentioned, Adobe was already a color management vendor and Adobe's experience became the ICC's wisdom. Adobe's main motivation was to support the PostScript concept that the same document printed at any time on any printer should always look the same. To make that happen, as much information as possible should be packed into the profile's required tags, with relatively little room for creativity in the image processing. This is consistent with a "smart profile" and a "dumb CMM," and this is how an ICC color management system currently works.

In June 1994 the first ICC profile format specification, version 3.0, was published.[5] Currently, version 4.2 is available. ICC profile format specifications are made available at the consortium website: www.color.org.

REFERENCES

1. IEC 61966-2-1 Amendment 1: 2003, Multimedia systems and equipment — Colour measurement and management — Part 2-1: Colour management — Default RGB colour space — sRGB, 2003.
2. ICC Specification ICC.1:2003-09 (Profile version 4.2.0.0) Image technology color management — Architecture, profile format, and data structure, 2004.
3. InterColor Consortium, InterColor Profile Format, Version 3.0, 1994.
4. Adobe Systems, Inc., *Postscript® Language Reference Manual* (2nd ed.), Addison-Wesley Professional, 1990.
5. J.E. Thornton, Y.J. Lee, and M.R. Balonon-Rosen, The Apple ColorSync Printer Profile Model and Its Optimization, *Proc. of IS&T's 46th Annual Conference*, pp. 147–150, 1993.

9 Desktop Spectral-Based Printing

*Mitchell R. Rosen, Francisco H. Imai,
Yongda Chen, Lawrence A. Taplin, and
Roy S. Berns*

CONTENTS

Although not ready for commercial use, an active area of research with significant potential impact on desktop printers is the innovation of spectral reproduction systems. Output from such systems will have more stable color relative to an original than is currently achieved. Spectral systems do not rely on colorimetry. Instead, they attempt to reproduce actual spectral reflectance or spectral transmittance of an original or its spectral radiance.

Advantages to these new systems will be many. A printed reproduction that matches the reflectance spectra of an original will preserve color matches over the range of all illuminants and for all observers. Reflectance-based rendering would be less sensitive to small characterization and calibration errors and to rendering noise. Radiance matching could be very powerful when reconstructing an object's appearance for viewing under new conditions. Proofing systems will improve as well as systems that make spot-color and specialty ink choices.

9.1 CURRENT METAMERIC SYSTEMS

In 1802, Young presented to the Royal Society his conclusion that human perception of color is based on three primaries. Young was correct. This phenomenon

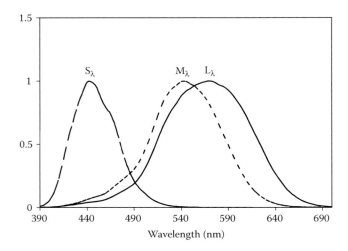

FIGURE 9.1 Spectral sensitivity of the three cone types.

is due to the typical character of sensors known as cones that populate the human retina. Cones are useful for seeing color under normal lighting conditions. There are three types of cones. S-cones have sensitivity primarily to the relatively short wavelengths, L-cones are biased toward relatively long wavelengths, and M-cones fall between them (see Figure 9.1). As long as the three cone types have the same responses when looking at two objects, the two objects will have the same color. Thus, three well-chosen primaries can stimulate the cones to see any color. The phosphors on a TV work in this way, as do cyan, magenta, and yellow inks in a printer.

Pairs of objects with distinct spectral reflectance properties but that arouse the same psychological color response when viewed under a single light source are known as metameric pairs. Metamerism has been important for color reproduction from the time of cave drawings to modern digital systems. In fact, until recently, almost all innovation in color imaging has been based on metamerism. Maxwell depended upon it for his 1861 demonstration of three-channel full-color photography and manufacturers of modern digital cameras, scanners, and printers still rely on it. Looking at the four major desktop printer technologies discussed in Part II of this book, it is clear that they, as currently used, create color through the use of metamerism. They depend primarily on the use of the standard process inks of cyan, magenta, and yellow to create the appearance of all colors.

9.2 TRADITIONAL VS. SPECTRAL-BASED SYSTEMS

In spite of its wide use, metameric reproduction has a number of drawbacks. Many are addressed by spectral color reproduction [Hunt, 1995; Berns, 1999; Hill, 2002]. Color from a printed copy will change its appearance under different illumination. If the original color is metameric with an original, it is likely that

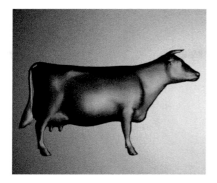 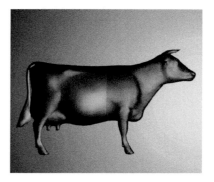

FIGURE 9.2 (See color insert following page 176.) Garrett Johnson's "Metameric Cows" [Johnson, 1998]. This demonstration simulates how the same reflectance properties under a single light source can appear different to different observers. The front half and the back half of the cow are different reflectances. To the two-degree observer (left), the front and the back have the same color. To the ten-degree observer (right), the front and the back have different colors.

under a different light source, the match between original and reproduction will break. There is also a surprisingly wide range of sensitivity differences among people. Thus, a pair of colors may appear metameric to one observer but not to another (see Figure 9.2).

Although the data flow diagram does not need to change between spectral and metameric systems, the details concerning each stage and the connections between stages are very different. Figure 9.3a shows a typical, contemporary, completely metameric system. In such a system, source and destination devices are characterized relative to colorimetry. This type of system was discussed in the previous chapter on color management. The input device captures three wideband red, green, blue (RGB) channels. The image is processed anticipating the colorimetric rendering capabilities of a cyan, magenta, yellow, black (CMYK) printer. Figure 9.3b shows the flow diagram for a spectral system. Here, spectral response and spectral rendering are characterized. The source device captures many narrowband channels. The image is processed minimizing spectral error for a many-colorant printer.

The remainder of the chapter is devoted to detailing spectral systems, showing their potential and limitations. Although spectral image acquisition and estimation are not the main emphases here, spectral printing systems will be put into the context of end-to-end scene-to-hardcopy reproduction.

9.3   SPECTRAL IMAGE ACQUISITION SYSTEM

As shown in Figure 9.3b, the spectral color printing system receives data captured by a multi-channel imaging system. The overall performance of the system will depend highly on the quality of the spectral information generated by the spectral image acquisition system. A variety of camera approaches is available for spectral

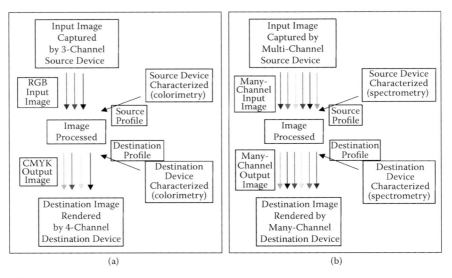

FIGURE 9.3 (a) Imaging flow diagram for a metameric color reproduction system. (b) Imaging flow diagram for a spectral color reproduction system.

image acquisition systems [Bacci et al., 1992; Bacci, 1995; Burns, 1997; König and Praefcke, 1998; Tominaga, 1999; Rosen and Jiang, 1999; 2002; Hauta-Kasari et al., 1999; Haneishi et al., 2000; Imai, Berns, and Tzeng, 2000a; Imai, Rosen, and Berns, 2000b; Imai et al., 2001; Murakami et al., 2001; Hardeberg, Schmitt, and Brettel, 2002]. In the early 1990s, Bacci et al. (1992, 1995) used a small-aperture optical fiber reflectance spectrophotomer to scan across painting and frescos. In general, there are two different image-capturing approaches for full-frame spectral images. The first is a narrowband approach [König and Praefcke, 1998; Tominaga, 1999; Rosen and Jiang, 1999; Imai, Rosen, and Berns, 2000b; Hardeberg, Schmitt, and Brettel, 2002]. Results have been reported on using a set of interference filters [König and Praefcke, 1998] or a narrow-bandpass tunable filter [Slawson, Ninkov, and Horch, 1999] in front of a monochrome sensor [Tominaga, 1999; Rosen and Jiang, 1999; Imai, Rosen, and Berns, 2000b; Hardeberg, Schmitt, and Brettel, 2002]. Both charge-coupled devices (CCDs) [Tominaga, 1999; Imai, Rosen and Berns, 2000b; Hardeberg, Schmitt, and Brettel, 2002] and panchromatic, black-and-white film [Rosen and Jianag, 1999] have been utilized as sensors for narrow-bandpass capture. Alternatively, Hauta-Kasari et al. used an optical system with a diffraction grating to produce spectrally adjustable illumination [Hauta-Kasari et al., 1999]. These systems are analogous to using a spectrophotometer, sampling the visible spectrum at known bandpass and wavelength intervals and, because they are spectral measurement methods, simple transformation can be generated from captured camera signals to reflectance spectra.

The second approach is an abridged technique that uses five or more absorption [Haneishi, et al., 2000; Imai, Rosen, and Berns, 2000b; Hardeberg, Schmitt, and Brettel, 2002] or interference [Burns, 1997] wideband filters in front of a monochromatic camera and requires spectral estimation [Marimont and Wandell, 1992]. The Visual Arts System for Archiving and Retrieval of Images project (VASARI) developed a system to capture paintings at high resolution [Saunders and Cupitt, 1993]. The VASARI scanner recorded the reflectance properties between 400 and 700 nm using 7 channels, each with a 70-nm bandwidth. Alternatively, it is possible to build a broadband rewritable filter [Miyazawa, Hauta-Kasari, and Toyooka, 2001].

In another approach, a conventional trichromatic digital camera is combined with selected absorption filters or light sources [Imai, 2000a; Sun, 2002]. The filters or light sources are selected to optimize the performance of the spectral estimation. In this broadband approach, the spectral reflectance of each pixel of the original scene can be estimated using *a priori* spectral analysis with direct spectrophotometric measurement and imaging of samples of the object to establish a relationship between the camera signals and spectral reflectance. The wideband acquisition takes advantage of the possibility of decreasing the spectral sampling increment without a significant loss of spectral information because of the smooth absorption characteristics of both manmade and natural colorants within the visible spectrum [Maloney, 1986; Parkkinen, 1989; Jaaskelainen, 1990; Dannemiller 1992; Vrhel 1992, 1994]. However, this technique has poorer performance compared to the previous methods because it is based on a color camera limited by its inherent spectral sensitivities. However, this method has the advantage of being more easily implemented.

The accuracy of the spectral estimation method can be evaluated by taking the similarity of the original and the estimated spectral curves into account or considering how the reproduced spectrum resembles the original scene when viewed by a human observer. Therefore, both spectral curve error metrics and colorimetric metrics have to be evaluated [Imai 2002a, 2002b, 2003a]. We recommend manually viewing spectral difference curves as an important aspect of system evaluation along with the calculation of multiple objective metrics such as color-difference equations, spectral curve difference metrics, and a metameric index calculation. A metameric index compares the extent to which two spectra are different between a reference condition and a test condition under different illuminants or observers [Nimeroff, 1965; Fairman, 1997; Chen et al., 2004]. In particular, the metameric index using Fairman parameric decomposition [Fairman, 1997] corrects the test spectrum until exact tristimulus equality is achieved under a reference condition. Then the metameric index is calculated using an International Commission on Illumination (CIE) color-difference equation for a test illuminant and observer. For the spectral curve difference metrics, we often use the root mean square error (rms) between original and estimated spectra and the goodness-of-fit coefficient (GFC) [Hernández-Andrés, 2001]. GFC is based on the inequality of Schwartz having values between 0 and 1 and indicates the correlation between two spectral curves; a value unity corresponds to a perfect spectral match. The metric is calculated using Equation 9.1:

$$GFC = \frac{\left| \sum_j R_m(\lambda_j) R_e(\lambda_j) \right|}{\sqrt{\left| \sum_j \left[R_m(\lambda_j) \right]^2 \right|} \sqrt{\left| \sum_j \left[R_e(\lambda_j) \right]^2 \right|}} \qquad (9.1)$$

where $R_m(\lambda_j)$ is the measured original spectral data at the wavelength λ_j and $R_e(\lambda_j)$ is the estimated spectral data at wavelength λ_j. According to the developers of the metric, GFC $\geq$ 0.999 and GFC $\geq$ 0.9999 are required for respectively good and excellent spectral matches [Hernández-Andrés, 2001].

Because we are considering complex images and not only single color stimulus, we also have to consider psychophysical evaluation after reproduction [Day, 2003].

The multi-band image acquisition and spectral estimation process can, by itself, produce a book involving many aspects such as estimation techniques [Mancill, 1975; Pratt, 1976; Praefke, 1996], encoding [Keusen, 1996], design and use of filters [Vent, 1994; Vrhel, 1995; Vora, 1997; Haneishi, 2000; Imai, 2001; Hardeberg, 2002; Rosen, 2002] and number of samples used in characterization [Tsumura, 1999].

Multi-band image acquisition and spectral estimation result in improvement over traditional image acquisition systems, even if the resulting image is rendered for a particular illuminant and all the subsequent color management is only colorimetric-based [Saunders, 1993; Imai, 1996; Sun, 2002; Day, 2003]. Spectral image acquisition can be a useful analytical tool for painting [Baronti, 1998; Casini, 1999; Berns, 2002] or human skin [Sun, 2002]. However, in reproduction terms, we can take full advantage of the spectral-based color image acquisition if this is followed by a subsequent spectral-based color image management and a spectra-based color separation printing.

9.4 SPECTRAL COLOR MANAGEMENT

In Chapter 8, traditional color management was discussed. Here we discuss how a spectral version of color management might be implemented. For a spectral input profile, Equation 8.1 and Equation 8.2 could easily be updated to a transformation such as found in Equation 9.2 and Equation 9.3, respectively.

$$[C_{n'}] = LUT_{C_n \to C_{n'}} [C_n] \quad (n:1 \text{ to } N) \qquad (9.2)$$

$$
\begin{bmatrix} f_1 \\ f_2 \\ \vdots \\ f_R \end{bmatrix} = \begin{bmatrix} a_{11} & a_{12} & \cdots & a_{1N} \\ a_{21} & a_{22} & \cdots & a_{2N} \\ \vdots & \vdots & & \vdots \\ a_{R1} & a_{R2} & \cdots & a_{RN} \end{bmatrix} \begin{bmatrix} C_{1'} \\ C_{2'} \\ \vdots \\ C_{N'} \end{bmatrix} \tag{9.3}
$$

where N is the number of input channels, C_n are the digital counts from the source device; a_{jii} is the matrix coefficient to estimate spectral reflectance or radiance; R is the number of samples per spectrum; and f_r is spectral reflectance or radiance from an object.

To update Equation 8.3 through Equation 8.5 for producing a spectral output, one is tempted to simply use analogous structures as those seen in the following equations:

$$
\left[f_{r'} \right] = LUT_{f_n \to f_{n'}} \left[f_r \right] \quad (\text{r: 1 to R}) \tag{9.4}
$$

$$
\begin{bmatrix} D_{1'} \\ D_{2'} \\ \vdots \\ D_{M'} \end{bmatrix} = LUT_{f_1 \dots R' \to D_1 \dots M'} \left[f_{1'}, f_{2'}, \cdots, f_{R'} \right] \tag{9.5}
$$

$$
\left[D_m \right] = LUT_{D_{m'} \to D_m} \left[D_{m'} \right] \quad (\text{m: 1 to M}) \tag{9.6}
$$

where M is the number of output channels and D_m are the digital counts for the destination device.

Equation 9.5 introduces a huge problem, however. One cannot apply a simple scale factor to all the image-processing approaches that work for metameric systems described by Figure 9.3a to create a system that works for the spectral systems of Figure 9.3b. When R in Equation 9.5 is sufficiently large, the multi-dimensional look-up table of that equation becomes far too large for implementation. Rosen, Ohta, and Derhak have attacked this problem through the introduction and development of the concept of the interim connection space (ICS) for spectral color management [Rosen, 2001a; 2003a; 2003b; Derhak, 2005].

Beyond changes in processing, the differences between data capture and throughput demands of the two systems described in Figure 9.3 motivate fresh approaches to imaging in a spectral system.

Discussion of spectral color management strategies has begun to gain some momentum within the community [Hung, 1999; Hill, 2000; Rosen, 2000, 2001a; 2001b; 2003a; 2003b; TAO, 2002]. The International Color Consortium (ICC) color management based on colorimetry shown in Figure 8.3 could be updated to spectral color management in a simple fashion, as shown in Figure 9.4.

Although the change from this viewpoint is quite simple, the tremendous increase in dimensionality for spectral processing will cause it to deviate dramatically from the ICC approach in its most efficient configuration.

In 2000, Rosen et al. described the basic aspects of a spectral profile. The device characterization found within a spectral profile needs a new profile connection space (PCS), one that may be based on reflectance, transmittance, or radiance. A color management system supporting spectral profiles will probably be required to know how to deal with any of these PCSs. Table 9.1 shows which spectral PCSs would be the most appropriate different device types.

Input profiles, within a spectral color management system, could describe any of the following transforms:

(a) Input digits to reflectance-based PCS or transmittance-based PCS
(b) Input digits to radiance-based PCS
(c) Input digits to a colorimetric PCS (not spectral, but still useful and essential for backward compatibility).

Color processing would need to be able to make the following conversions:

(d) Reflectance or transmittance to radiance by multiplying an illuminant
(e) Radiance to reflectance or transmittance by dividing an illuminant
(f) Reflectance or transmittance to colorimetry by multiplying an illuminant, multiplying color-matching functions, and then integrating
(g) Radiance to colorimetry by multiplying color-matching functions and then integrating.

A destination profile could describe any of the following transforms:

(h) Reflectance-based PCS or transmittance-based PCS to output digits
(i) Radiance-based PCS to output digits
(j) Colorimetric-based PCS to output digits

These basic operations could be applied in series to carry out complex tasks. For example, consider the problem of choosing the right wallpaper for one's tungsten-lit living room. It is well known that the narrowband fluorescents in the store can be misleading. A customer sophisticated in spectral reproduction techniques could take out a multispectral camera, capture a picture of the wallpaper in the store, and take a second shot of the store's lights. At home the two images could be downloaded to a computer. Using the camera's spectral profile, each image would be converted to radiance (functionality (b), above). Using the radiance image of the store light source, the wallpaper radiance image could then be transformed to reflectance (functionality (e), above). A subsequent picture of the living room light would allow for calculation of what the wallpaper would have looked like at home (functionality (d), above). Finally, to view the color image on the home monitor, the radiance image would be converted to XYZ

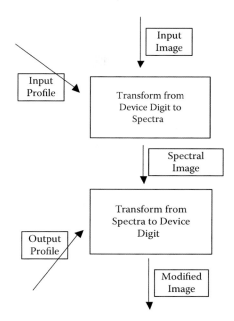

FIGURE 9.4 Possible logical spectral image-processing block diagram.

TABLE 9.1
Spectral PCS Basis for Device Types

Device Type	Spectral PCS Basis	Alternative Spectral PCS Basis
Scanner	Reflectance or Transmittance	
Camera	Radiance	Reflectance (for illuminant-controlled studio capture)
Display	Radiance	
Printer	Reflectance or Transmittance	Radiance when viewing conditions are defined

FIGURE 9.5 (See color insert following page 176.) Spectral color management would provide a way to impose a new light source on the image of a captured object.

(functionality (g), above) and then the ICC monitor profile would be used to transform to monitor RGB (functionality (j), above). Figure 9.5 illustrates a cartoon of the user action in this example. Figure 9.6 shows the series of actions taken by the spectral color management system.

The most significant work done to date for defining a spectral profile has emerged from the Akasaka Natural Vision Research Center of the Telecommunications Advancement Organization (TAO) of Japan. In its 2002 research report [TAO, 2002] a detailed proposal for many aspects of a spectral profile, was described. Borrowing data structures and format from the ICC specification, the Natural Vision group put together an important first step.

The Akasaka report included some of the important structures that a spectral profile would need. Importantly missing from the report, however, was a description of a profile tag for the spectral characterization of a printing device. Multichannel input devices and multi-primary displays were included. A spectral printer, however, brings up issues of characterization and processing that far exceed the considerations for the previous tags.

The Natural Vision proposal included two spectral PCSs. Both were defined between 380 and 780 nm in 1-nm increments. The first space was specified in units of W/sr/m^2/nm and the other was specified in units of reflectance factor. Following the logic of Table 9.1, most cameras and display devices could use the first of these PCSs. Most scanners and printers could use the second one.

Three profile types important to spectral color management were defined in the Akasaka report. The N-component spectrum-based input profile and the N-component spectral matrix-based input profile both contained tags supporting transformation described by Equation 9.7 and Equation 9.8. The M-primary display profile supports the transformation described by Equation 9.9 and Equation 9.10.

$$\text{linear}C_n = mTRC[C_n - b_n] \ (n: 1 \text{ to } N) \tag{9.7}$$

$$
\begin{bmatrix} f_1 \\ f_2 \\ \vdots \\ f_R \end{bmatrix}
=
\begin{bmatrix}
\dfrac{1}{k_n} & 0 & \cdots & 0 \\
0 & \dfrac{1}{k_n} & \cdots & 0 \\
\vdots & & & \vdots \\
0 & \cdots & 0 & \dfrac{1}{k_n}
\end{bmatrix}
\begin{bmatrix}
a_{11} & a_{12} & \cdots & a_{1N} \\
a_{21} & a_{22} & \cdots & a_{2N} \\
\vdots & \vdots & & \vdots \\
a_{R1} & a_{R2} & \cdots & a_{RN}
\end{bmatrix}
\begin{bmatrix} \text{linear}\,C_1 \\ \text{linear}\,C_2 \\ \vdots \\ \text{linear}\,C_N \end{bmatrix}
\tag{9.8}
$$

where b_n is the bias digital count; C_n is the digital count of the input device (0–1); a_{ij} are the matrix coefficients to estimate spectral reflectance or radiance; k_n is the coefficient of sensitivity-level correction; R is 401; and f_R is spectral reflectance or radiance from an object.

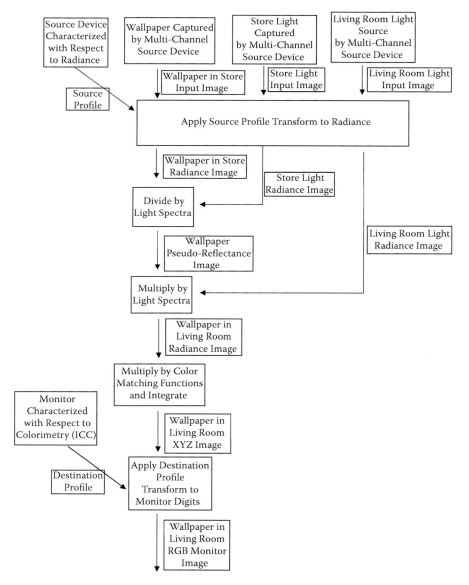

FIGURE 9.6 Block diagram illustrating the logical color management system actions that would accompany the wallpaper example described.

$$\text{linearC}_n = \text{mTRC}[\text{C}_n] \quad (n: 1 \text{ to } N) \tag{9.9}$$

$$\begin{bmatrix} r_1 \\ r_2 \\ \vdots \\ r_R \end{bmatrix} = \begin{bmatrix} p_{11} & p_{12} & \cdots & p_{1N} \\ p_{21} & p_{22} & \cdots & p_{2N} \\ \vdots & \vdots & & \vdots \\ p_{R1} & p_{R2} & \cdots & p_{RN} \end{bmatrix} \begin{bmatrix} \text{linear C}_1 \\ \text{linear C}_2 \\ \vdots \\ \text{linear C}_N \end{bmatrix} + \begin{bmatrix} p_1 \\ p_2 \\ \vdots \\ p_R \end{bmatrix} \tag{9.10}$$

where C_n is the digital count of the input device (0–1); p_{RN} is the spectral radiance of the nth primary measured at maximum digital count; $p_{n,r}$ is the bias spectrum; R is 401; and r_R is the radiance of an object on the display.

In the spirit of the spectral color management systems previously described by Rosen [Rosen, 2000, 2001a], Yamaguchi and coworkers in the TAO report illustrated the use of the spectral profiles in color management systems that support both spectral and colorimetric profiles [Yamaguchi, 2002].

9.5   SPECTRAL MODEL FOR PRINTERS

The Cellular–Yule–Nielsen–Spectral–Neugebauer (CYNSN) model [Wyble, 2000] was used to investigate a spectral model for printers. The Neugebauer model [Neugebauer, 1937] is an additive model for multi-ink printing in which a macroscopic colored area is a weighted sum of the individual microscopic colors. The weights are determined from the halftoning algorithm. Because of light scattering within the paper, the relationship between the macroscopic and microscopic colors becomes complicated. Yule and Nielsen [Yule, 1951] found that exponentiating reflectance, in similar fashion to converting reflectance to optical density, greatly improved prediction accuracy. Viggiano [Viggiano, 1985] further improved performance by considering the optical mixing over a narrow range of wavelengths. The resulting Yule–Nielsen–Spectral–Neugebauer (YNSN) model is shown below:

$$R_\lambda = \left(\sum_i F_i R_{\lambda,i}^{1/n} \right)^n \tag{9.11}$$

where $R_{\lambda,i}$ is the macroscopic spectral reflectance of the ith color type at 100% area coverage, n is the Yule–Nielsen exponent, and F_i are the fractional area coverages of each microscopic color type. The maximum value of i depends on the number of inks and the halftoning algorithm. For three-color printing that uses rotated screens or frequency modulation, the maximum number is eight (e.g., cyan, magenta, yellow, red, green, blue, black, and paper white). That is, three inks printed randomly result in eight unique colors; for six-color printing, the result is 64 colors. These colors are known as the Neugebauer primaries. The fractional areas are determined as a product of random variables, shown in

Equation 9.12. These probabilities when used for printing are attributed to Demichel [Demichel, 1924].

$$F_i = \prod_j \begin{pmatrix} a_j & \text{if ink j is in Neugebauer Primary i} \\ (1 - a_j) & \text{if ink j is not in Neugebauer Primary i} \end{pmatrix} \quad (9.12)$$

where a_j is the effective area coverage of ink j. (The term *effective* is used because this area coverage is determined statistically, not optically using reflection microscopy [Wyble, 2000].) Area coverage is a function of the digital signal, d_j, controlling the amount of ink delivered to the substrate, defined in Equation 9.13.

$$a_j = f(d_j) \quad (9.13)$$

The n value and the effective area coverage relationships are determined statistically, typically using one-color ramps.

From a geometric viewpoint, the YNSN model performs multi-dimensional linear interpolation across $R_\lambda^{1/n}$ space, the interpolation weights calculated using Equation 9.12 [Rolleston, 1993; Balasubramanian, 1999]. Heuberger [1992] recognized that interpolation performance is always improved by reducing the interpolation area. This is easily achieved by creating subspaces called cells. Rolleston and Balasubrananian evaluated the cellular method when using the YNSN model for printer characterization, creating the Cellular–Yule–Nielsen–Spectral–Neugebauer or CYNSN model. Improvement was shown to be significant. In particular, the cellular approach greatly reduced the need for highly accurate analytical models beyond what was typically achieved using the YNSN model.

The value of the cellular approach is shown graphically in Figure 9.7. Here, the outer square and solid circles represent the YNSN model, and the point O1 is calculated by interpolation from four outer corner points, P1, P2, P3, and P4, which are represented by the solid circles, the Neugebauer primaries. The whole figure, including solid and dashed circles, represents the CYNSN model in two dimensions. If each ink is printed at four levels, there are more known values and can be used to create cells (subspaces). The corners of each cell are the cellular Neugebauer primaries, or simply cellular primaries. If the cellular model is used to predict the point O1, we can use the nearest four cellular primaries (P11, P22, P33, and P44). The accuracy improvement of the cellular model is significant because interpolation is performed in a much smaller subspace. Of course, the cost is that more colors need to be printed and measured. Agar and Allebach [Agar, 1998] showed the relationship between prediction error and number of cellular primaries. The accuracy of the cellular model can be improved significantly as more primaries are considered, though, as noted by Balasubramanian, there is a diminishing return.

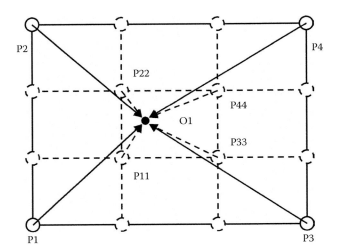

FIGURE 9.7 Graphical interpretation of the YNSN and CYNSN models.

Iino and Berns [1998a, 1998b] used the YNSN model to characterize an inkjet printer and a proofing system for offset printing with good success. Balasubramanian [Balasubramanian, 1996, 1999] evaluated the spectral Neugebauer (SN), YNSN, and CYNSN models when characterizing a four-color electrophotographic printer. He used a typical black-printer strategy so that the number of samples required to create a four-color model was not excessive. As expected, accuracy improved significantly by adding the Yule–Nielsen n value. The addition of the cellular subspaces resulted in modest incremental improvement. Balasubramanian also investigated using weighted linear regression to optimize the spectral properties of the Neugebauer primaries. Rather than using macroscopic measurements, he optimized the spectral properties of those primaries not containing black ink resulting in the best average colorimetric performance. Samples near the primary in colorant space were weighted more heavily than samples far away. The improvement to the YNSN model was similar to adding the cellular approach. Most recently, Imai, Wyble, and Tzeng [Imai et al., 2003b] used the YNSN model to characterize a CMYK inkjet printer as part of an end-to-end spectral color reproduction system with reasonable success.

Tzeng [1999a] and Tzeng and Berns [2000] used the YNSN model for six-color proofing using cyan, magenta, yellow, black, orange, and green inks. One goal was to use the proofer to simulate different ink sets. Accordingly, a spectral model was desired so that proofs could be produced that were minimally metameric. That is, a printing system defined spectrally either computationally or by direct measurement could be proofed via spectral color reproduction. This required numerically inverting the YNSN model. In order to ensure convergence and reduce processing time, Tzeng subdivided the six inks into 10 four-ink models. For any color, only four inks would be printed. Model accuracy depended on the particular four inks. Taplin and Berns [Taplin, 2001] extended Tzeng's research and considered all six inks simultaneously. They used an inkjet printer with a small dot size, an error diffusion halftoning algorithm, and heavyweight art paper. This combination enabled all of the 64 Neugebauer primaries to be printed without ink blotting. The YNSN model was used again with reasonable performance accuracy. More sophisticated optimization algorithms were used for model inversion as well as a continuous-tone model to provide reasonable starting values. The printer was used for a spectral color reproduction system for artwork [Art-SI, 2004].

9.6 CONCLUSIONS

Today, there are specialized areas where capturing and reproducing the spectra of an original scene or document are considered important goals. These include reproduction and archiving of artwork, proofing, medical imaging, remote sensing, and catalog sales. Many other potential opportunities exist for the use of spectral information in a color reproduction workflow. Given the groundswell of interest and investigation into spectral imaging capabilities, it is likely that developments will accelerate quickly. Some consumer camera systems are already increasing the number of input channels, and many desktop printers have six to eight inks. There is every reason to believe that spectral reproduction will move to the desktop as the technology continues to mature.

ACKNOWLEDGMENT

Section 9.5 is reprinted from [Chen, 2004], with permission from IS&T: The Society for Imaging Science and Technology, sole copyright owners of *The Journal of Imaging Science and Technology.*

REFERENCES AND BIBLIOGRAPHY

A.U. Agar and J.P. Allebach, An interative cellular YNSN method for color printer characterization, *Proc. IS&T/SID Sixth Color Imaging Conference*, pp. 197–200 (1998).

J.S. Arney, A probability description of the Yule–Nielsen effect I, *J. Imag. Sci. Technol.* 41, pp. 633–636 (1997a).

J.S. Arney, A probability description of the Yule–Nielsen effect II: The impact of halftone geometry, *J. Imag. Sci. Technol.* 41, pp. 637–642 (1997b).

J.S. Arney, C.D. Arney, and P.G. Engeldrum, Modeling the Yule–Nielsen effect, *J. Imag. Sci. Technol.* 40, pp. 233–238 (1996).

Art-SI, www.art-si.org, Art Spectral Imaging (2004).

M. Bacci, Fibre optics applications to works of art, *Sensors and Actuators B* 29, pp. 190–196 (1995).

M. Bacci, S. Baronti, A. Casini, F. Lotti, M. Picollo, and O. Casazza, Non-destructive spectroscopic investigations on paintings using optical fibers, *Proc. Materials Res. Soc. Symp.* 267, pp. 265–283 (1992).

R. Balasubramanian, The use of spectral regression in modeling halftone color printers, *Proc. of IS&T/OSA Optics and Imaging in the Information Age*, pp. 372–375 (1996).

R. Balasubramanian, Optimization of the spectral Neugebauer model for printer characterization, *J. Electronic Imaging* 8, pp. 156–166 (1999).

S. Baronti, A. Casini, F. Lotti, and S. Porcinai, Multispectral imaging system for the mapping of pigments in works of art by use of principal-component analysis, *Appl. Optics* 37, pp. 1299–1309 (1998).

R.S. Berns, Spectral modeling of a dye diffusion thermal transfer printer, *J. Electronic Imaging* 2, pp. 359–370 (1993).

R.S. Berns, Challenges for colour science in multimedia imaging systems, in L. MacDonald and M.R. Luo, Eds., *Colour Imaging: Vision and Technology*, John Wiley & Sons, England, pp. 99–127 (1999).

R.S. Berns, *Billmeyer and Saltzman's Principles of Color Technology*, 3rd ed., John Wiley & Sons (2000).

R.S. Berns and M. Shyu, Colorimetric characterization of a desktop drum scanner using a spectral model, *J. Electronic Imaging* 4, pp. 360–372 (1995).

R.S. Berns, J. Krueger, and M. Swicklik, Multiple pigment selection for inpainting using visible reflectance spectrophotometry, *Studies in Conservation* 47, pp. 46–61 (2002).

H. Boll, A color to colorant transformation for a seven ink process, *Proc. IS&T Third Technical Sympos. Prepress Proofing Printing*, pp. 31–36 (1993).

P.D. Burns, Analysis of Image Noise in Multi-Spectral Color Acquisition, Ph.D. Thesis, R. I. T., Rochester, NY (1997).

A. Casini, F. Lotti, M. Picollo, L. Stefani, and E. Buzzegoli, Image spectroscopy mapping technique for non-invasive analysis of paintings, *Studies in Conservation* 44, pp. 39–48 (1999).

Y. Chen, R.S. Berns, and L.A. Taplin, Six color printer characterization using an optimized cellular Yule–Nielsen spectral Neugebauer model, *J. Imaging Technol.*, 48, pp. 519–528 (2004).

J.L. Dannemiller, Spectral reflectance of natural objects: how many basis functions are necessary?, *J. Opt. Soc. Am.* A9, pp. 507–515 (1992).

E.A. Day, The Effects of Multi-Channel Spectrum Imaging on Perceived Spatial Image Quality and Color Reproduction Accuracy, M.S. Thesis, R. I.T., Rochester, NY (2003).

M.E. Demichel, *Procédé* 26 pp. 17–21, 26–27 (1924).

M.W. Derhak and M.R. Rosen, Spectral colorimetry using LabPQR — an interim connection space, J. Imaging Sci. Technol., in press.

P. Emmel and R.D. Hersch, A unified model for color prediction of halftoned prints, *J. Imaging Sci. Technol.* 44, pp. 351–359 (2000).

H.S. Fairman, Metameric correction using parametric decomposition, *Color Res. Appl.* 12, pp. 261–265 (1997).

H. Haneishi, T. Suzuki, N. Shimoyama, and Y. Miyake, Color digital halftoning taking colorimetric color reproduction into account, *J. Electronic Imaging* 5, pp. 97–106 (1996).

H. Haneishi, T. Hasegawa, A. Hosoi, Y. Yokohama, N. Tsumura, and Y. Miyake, System design for accurately estimating the spectral reflectance of art paintings, *Appl. Opt.* 39, pp. 6621–6632 (2000).

W. Hanson, Color photography: from dream, to reality, to commonplace, in E. Ostroff, Ed., *Pioneers of Photography, Their Achievements in Science and Technology*, SPSE, Springfield (1986).

J.Y. Hardeberg, F. Schmitt, and H. Brettel, Multispectral color image capture using a liquid crystal tunable filter, *Optical Engineering* 41, pp. 2533–2548 (2002).

M. Hauta-Kasari, K. Miyazawa, S. Toyooka, and J. Parkkinen, Spectral vision system for measuring color images, *J. Opt. Soc. Am.* A16, pp. 2352–2362 (1999).

R. Herbert, Hexachrome color selection and separation — model for print media, *Proc. IS&T 3rd Technical Sympos. Prepress Proofing and Printing*, pp. 28–30 (1993).

J. Hernández-Andrés, J. Romero, J. L. Nieves, and R. L. Lee Jr., Color and spectral analysis of daylight in southern Europe, *J. Opt. Soc. Am.* A18, pp. 1325–1335 (2001).

K.J. Heuberger, Z.M. Jing, and S. Persiev, Color transformations and lookup tables, *Proc. TAGA/ISCC*, pp. 863–881, (1992).

B. Hill, Color capture, color management and the problem of metamerism: does multispectral imaging offer the solution? *Proceedings of SPIE*, 3963, pp. 2–14 (2000).

B. Hill, (R)evolution of color imaging systems, *Proceedings First Europ. Conf. Color Graphics, Imaging Vision*, pp. 473–479 (2002).

P.C. Hung, Color reproduction using spectral characterization, *Proc. Intl. Sympos. Multispectral Imaging Color Reproduction Digital Arch.*, pp. 98–105 (1999).

P.C. Hung, T. Mitsuhashi, and T. Saitoh, Inkjet printing system for textile using Hi-fi colors, *Proc. PICS Conference*, pp. 46–50 (2001).

R.W.G. Hunt, *The Reproduction of Colour*, 5th ed., Fountain Press, Kingston-upon-Thames, U.K. (1995).

K. Iino and R.S. Berns, Building color management modules using linear optimization I. Desktop color system, *J. Imaging Sci. Tech.* 42, pp. 79–94 (1998a).

K. Iino and R.S. Berns, Building color management modules using linear optimization II. Prepress system for offset printing, *J. Imaging Sci. Tech.* 42, pp. 99–144 (1998b).

F.H. Imai, N. Tsumura, H. Haneishi, and Y. Miyake, Principal component analysis of skin color and its application to colorimetric color reproduction on CRT display and hardcopy, *J. Imaging Sci. Tech.* 40, pp. 422–429 (1996).

F.H. Imai, Multi-Spectral Image Acquisition and Spectral Reconstruction using a Trichromatic Digital Camera System Associated with Absorption Filters, *MCSL Technical Report* (1998).

F.H. Imai, R.S. Berns, and D. Tzeng, A comparative analysis of spectral reflectance estimation in various spaces using a trichromatic camera system, *J. Imaging Sci. Technol.* 44, pp. 280–287 (2000a).

F.H. Imai, M.R. Rosen, and R.S. Berns, Comparison of spectrally narrow-band capture versus wideband with a priori sample analysis for spectral reflectance estimation, *Proc. Eighth Color Imaging Conf.*, pp. 234–241 (2000b).

F.H. Imai, S. Quan, M.R. Rosen, and R.S. Berns, Digital camera filter design for colorimetric and spectral accuracy, *Proc. Third Intl. Conf, Multispectral Color Sci.*, University of Joensuu, Finland, pp. 13–16 (2001).

F.H. Imai, M.R. Rosen, and R.S. Berns, Comparative study of metrics for spectral match quality, *Proc. IS&T's First European Conf. Color Graphics, Imaging Vision*, pp. 492–496 (2002a).

F.H. Imai, L.A. Taplin, and E.A. Day, Comparison of the Accuracy of Various Transformations from Multi-Band Images to Reflectance Spectra, MCSL Technical Report (2002b).

F.H. Imai, L.A. Taplin, and E.A. Day, Comparative Study of Spectral Reflectance Estimation Based on Broadband Imaging Systems, MCSL Technical Report (2003a).

F.H. Imai, D.R. Wyble, and D. Tzeng, A feasibility study of spectral color reproduction, *J Imag Sci Tech* 47, p. 543 (2003b).

T. Jaaskelainen, J. Parkkinen, and S. Toyooka, Vector-subspace model for color representation, *J. Opt. Soc. Am. A*7, pp. 725–730 (1990).

G.M. Johnson, Computer Synthesis of Spectroradiometric Images for Color Imaging Systems Analysis, M.S. Thesis, R.I.T., Rochester, NY, (1998).

T. Keusen, Multispectral color system with an encoding format compatible with the conventional tristimulus model, *J. Imaging Sci. Tech.* 40, pp. 510–515 (1996).

T. Kohler and R.S. Berns, Reducing metamerism and increasing gamut using five or more colored inks, *Proc. IS&T Third Technical Sympos. Prepress, Proofing and Printing*, pp. 24–28 (1993).

F. König and W. Praefcke, A multispectral scanner, in L. MacDonald and M.R. Luo, Eds., *Colour Imaging: Vision and Technology*, John Wiley & Sons, Chichester, pp. 129–144 (1998).

P. Kubelka, New contribution to the optics of intensely light-scattering materials. Part I, *J. Opt. Soc. Am.* 38, pp. 448–457 (1948).

H. Kueppers, Process for manufacturing systematic color tables or color charts for seven-color printing, and tables or charts produced by this process, U.S. patent number 4878977 (1989).

L.T. Maloney and B.A. Wandell, Color constancy: a method for recovering surface spectral reflectance, *J. Opt. Soc. Am. A*3, pp. 29–33 (1986).

C.E. Mancill, Digital Color Image Restoration, Ph.D. Thesis, University of Southern California, Los Angeles (1975).

D.H. Marimont and B.A. Wandell, Linear models of surface and illuminant spectra, *J. Opt. Soc. Am. A* 9, pp. 1905–1913 (1992).

K. Miyazawa, M. Hauta-Kasari, and S. Toyooka, Rewritable broadband color filters for spectral image analysis, *Optical Review* 8, pp. 112–119 (2001).

Y. Murakami, T. Obi, M. Yamaguchi, N. Ohyama, and Y. Komiya, Spectral reflectance estimation from multi-band image using color chart, *Opt. Commun.* 188, pp. 47–57 (2001).

H.E.J. Neugebauer, Die theoretischen grundlagen des mehrfarbendrucks, Zeitscrift fur wissenschaftliche Photographie [Reprinted in *Proc. SPIE: Neugebauer Memorial Seminar on Color Reproduction* 1184, pp. 194–202 (1989)], (1937).

I. Nimeroff and J.A. Yurow, Degree of metamerism, *J. Opt. Soc. Am.* 55, pp. 185–190 (1965).

N. Ohta, Structure of the color solid obtainable by three subtractive color dyes, *Die Farbe* 20, pp. 115–134 (1971).

N. Ohta, The color gamut obtainable by the combination of subtractive color dyes IV. Influence of some practical constraints, *Photograph. Sci. Eng.* 28, pp. 228–231 (1982).

V. Ostromoukhov, Chromaticity gamut enhancement by heptatone multi-color printing, *Proc. of SPIE* 1909, pp. 139–151 (1993).

J. Parkkinen, J. Hallikainen, and T. Jaaskelainen, Characteristic spectra of Munsell colors, *J. Opt. Soc. Am. A* 4, pp. 318–322 (1989).

W. Praefcke, Transform coding of reflectance spectra using smooth basis vectors, *J. Imaging Sci. Tech.* 40, pp. 543–548 (1996).

W.K. Pratt and C.E. Mancill, Spectral estimation techniques for the spectral calibration of a color image scanner, *Appl. Opt.* 15, pp. 73–75 (1976).

R. Rolleston and R. Balasubramanian. Accuracy of various types of Neugebauer model, *Proc. Color Imaging Conference,* pp. 32–37 (1993).

M.R. Rosen, Navigating the Roadblocks to Spectral Color Reproduction: Data-Efficient Multi-Channel Imaging and Spectral Color Management, Ph.D. Dissertation, RIT, 2003a.

M.R. Rosen and N. Ohta, Spectral color processing using an interim connection space, Proc. 11th CIC, 187–192, 2003b.

M.R. Rosen and X. Jiang, Lippmann 2000: A spectral image database under construction, *Proc. International Sympos. Multispectral Imaging Color Reprod. Digital Arch.*, Chiba University, Chiba, Japan, pp. 117–122 (1999).

M.R. Rosen, E.F. Hattenberger, and N. Ohta, Spectral redundancy in a six-ink jet printer, *J. Imaging Sci. Technol.*, 48, 192–202, 2004.

M.R. Rosen, M.D. Fairchild, G.M. Johnson, and D.R. Wyble, Color management within a spectral image visualization tool, *Proc. Eighth Color Imaging Conf.* pp. 75–80 (2000).

M.R. Rosen, F.H. Imai, X. Jiang, and N. Ohta, Spectral reproduction from scene to hardcopy II: image processing, *Proc. of SPIE* 4300, pp. 33–41 (2001a).

M.R. Rosen, L.A. Taplin, F.H. Imai, R.S. Berns, and N. Ohta, Answering Hunt's web shopping challenge: spectral color management for a virtual swatch, *Proc. Ninth Color Imaging Conf.,* pp. 267–273 (2001b).

M.R. Rosen, F.H. Imai, M.D. Fairchild, and N. Ohta, Data-efficient methods applied to spectral image capture, *J. Soc. Photogr. Sci. Technol. Japan* 65, pp. 353–362 (2002).

T. Sato, Y. Nakano, T. Iga, S. Nakauchi, and S. Usui, Color reproduction based on low dimensional spectral reflectance using the principal component analysis, *Proc. IS&T/SID Fourth Color Imaging Conf.,* pp. 185–188 (1996).

D. Saunders and J. Cupitt, Image processing at the National Gallery: the VASARI project, *National Gallery Tech. Bull.* 14, pp. 72–86 (1993).

L. Sipley, *A Half Century of Color*, Macmillan, New York (1951).

L. Sipley, *Photography's Great Inventors*, American Museum of Photography, Philadelphia (1965).

R.W. Slawson, Z. Ninkov, and E.P. Horch, Hyperspectral imaging: wide-area spectrophotometry using a liquid-crystal tunable filter, *Publ. Astronomical Society Pacific* 111, pp. 621–626 (1999).

E.J. Stollnitz, V. Ostromoukhov, and D.H. Salesin, Reproducing color images using custom inks, *Computer Graphics Proc., Annu. Conf. Ser.*, pp. 267–274 (1998).

Q. Sun and M.D. Fairchild, Statistical characterization of face spectral reflectances and its application to human portraiture spectral estimation, *J. Imaging Sci. Technol.* 46, pp. 498–506 (2002).

L.A. Taplin and R.S. Berns, Spectral color reproduction based on a six-color inkjet output system, *Proc. Ninth Color Imaging Conf.*, pp. 209–213 (2001).

Telecommunications Advancement Organization of Japan, R&D Report on Image Presentation and Transmission System for Next Generation, pp. 22–79, *in Japanese* (2002).

S. Tominaga, Spectral imaging by a multichannel camera, *J. Electronic Imaging* 8, pp. 332–341 (1999).

N. Tsumura, H. Sato, T. Hasegawa, H. Haneishi, and Y. Miyake, Limitations of color samples for spectral estimation from sensor responses in fine art painting, *Optical Rev.* 6, pp. 67–61 (1999).

D. Tzeng, Spectral-Based Color Separation Algorithm Development for Multiple-Ink Color Reproduction, Ph.D. Thesis, R. I.T., Rochester, NY (1999).

D. Tzeng and R.S. Berns, Spectral-based ink selection for multiple-ink printing I. Colorant estimation of original objects, *Proc. IS&T/SID Sixth Color Imaging Conf.*, IS&T, Springfield, VA, pp. 106–111 (1998a).

D. Tzeng and R.S. Berns, Spectral reflectance prediction of ink overprints by Kubelka-Munk turbid media theory, *Proc. TAGA/ISCC Sympos.*, pp. 682–697 (1999b).

D. Tzeng and R.S. Berns, Spectral-based ink selection for multiple-ink printing II. Optimal ink selection, *Proc. IS&T/SID Seventh Color Imaging Conf.*, pp. 182–187 (1999c).

D. Tzeng and R.S. Berns, Spectral-based six-color separation minimizing metamerism, *Proc. IS&T SID Eighth Color Imaging Conf.*, pp. 342–247 (2000).

D.S. Vent, Multichannel Analysis of Object-Color Spectra, M.S. Thesis, R.I.T., Rochester, NY (1994).

J.A.S. Viggiano, The color of halftone tints, *Proc. TAGA* 37, pp. 647 (1985).

P.L. Vora and H.J. Trussell, Mathematical methods for the design of color scanning filters, *IEEE Trans. Image Processing* 6, pp. 312–320 (1997).

M.J. Vrhel and H.J. Trussell, Color correction using principal components, *Color Res. Appl.* 17, pp. 328–338 (1992).

M.J. Vrhel and H.J. Trussell, Optimal color filters in the presence of noise, *IEEE Trans. Image Processing* 4, pp. 814–823 (1995).

M.J. Vrhel, R. Gershon, and L.S. Iwan, Measurement and analysis of object reflectance spectra, *Color Res. Appl.* 19, pp. 4–9 (1994).

D.R. Wyble and R.S. Berns, A critical review of spectral models applied to binary color printing, *Color Res. Appl.* 25, pp. 5–19 (2000).

M. Yamaguchi, T. Taraji, K. Ohsawa, T. Uchiyama, H. Motomura, Y. Murakami, and N. Ohyama, Color image reproduction based on the multispectral and multiprimary imaging: experimental evaluation, *Proc. SPIE*, 4663, pp. 15–26 (2002).

J.A.C.Yule and W.J. Nielsen, The penetration of light into paper and its effect on halftone reproductions, *Proc. TAGA* 3, pp. 65 (1951).

S. Zuffi, R. Schettini, and G. Mauri, Using genetic algorithms for spectral-based printer characterization, *Proc. SPIE* 5008, pp. 268–275 (2003).

Index